{ PRAIRIE }

CANDACE SAVAGE

PRAIRIE

A NATURAL HISTORY

{ updated, with a new preface }

PRINCIPAL PHOTOGRAPHY BY JAMES R. PAGE

ILLUSTRATIONS BY JOAN A. WILLIAMS

 David Suzuki Foundation

GREYSTONE BOOKS
D&M PUBLISHERS INC.
Vancouver/Toronto/Berkeley

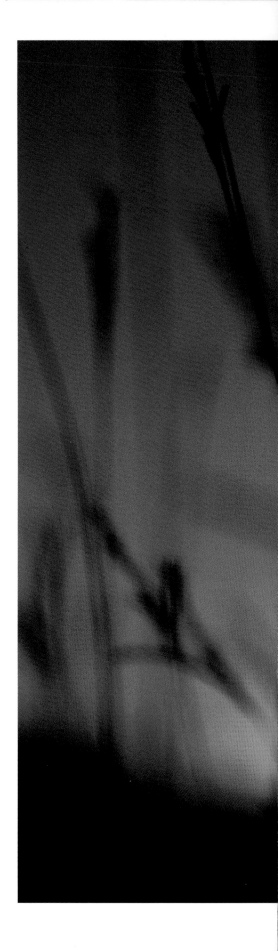

To the memory of my parents
Harry Sherk, 1920–2003, and Edna Sherk, 1919–2008

Greystone Books
An imprint of D&M Publishers Inc.
2323 Quebec Street, Suite 201
Vancouver BC Canada V5T 4S7
www.greystonebooks.com

David Suzuki Foundation
2211 West 4th Avenue, Suite 201
Vancouver BC Canada V6K 4S2

Cataloguing data available from Library and Archives Canada
ISBN 978-1-55365-588-6 (pbk.)

Editing by Nancy Flight
Copy editing by Barbara Tomlin and Lara Kordic
Cover and interior design by Naomi MacDougall
Front cover photograph by Joel Sartore/
National Geographic/Getty Images
Cartography by the Canadian Plains Research Center
Printed and bound in China by C&C Offset Printing Co., Ltd.
Text printed on acid-free paper
Distributed in the U.S. by Publishers Group West

We gratefully acknowledge the financial support of the Canada
Council for the Arts, the British Columbia Arts Council,
the Province of British Columbia through the Book Publishing
Tax Credit, and the Government of Canada through
the Canada Book Fund for our publishing activities.

MIX
Paper
FSC FSC™ C008047

CONTENTS

There is no way to hold back the future.
But we can shape the course of events by
engaging—fully, deeply, and passionately—
with the present . . . This approach is
sometimes referred to as a strategy of "no regrets,"
because the work is worth doing now,
no matter what happens next.

EVEN NOW, SEVEN years after the fact, I can vividly recall the moment when I wrote those words, read them back to myself, and realized that I was done. My book on grassland ecology and conservation, the impossible project that had occupied me night and day for so many years, was finally finished. At the time, my main emotion was not so much elation—the satisfaction of a job well done—as giddy relief that I had managed to get the thing completed, somehow. It hadn't been easy. Just as I sat down to write the concluding chapter, my partner, Keith, was diagnosed with advanced prostate cancer. (Don't worry: he's alive and well.) A couple of weeks later, my father suffered a stroke and died in hospital.

No matter what happens next. Being alive is a risky business, and the inevitable conclusion of our life stories is not what, given our druthers, most of us would choose. We're born; we die. And between the time when the lights switch on and the lights switch off, what are we to do? Let's assume that you and I number among the fortunate minority of humanity who enjoy reasonable access to the basic necessities: food, clothing, shelter, and community. With our survival needs met, how do we "improve each shining hour" so that our brief lives are not a flash in the pan but a flash of brilliance? How do we craft lives of purpose and significance?

...*because the work is worth doing*...These questions lurk, like ephemeral companions, at the edge of our field of vision. For me, the answers have often turned out to be comically understated: growing carrots in my garden, playing fiddle tunes on my accordion, or stringing words one after another to form sentences. But none of these activities can compare with the simple, animal pleasure of scuffing across a dusty stretch of half-wild prairie somewhere in the back of beyond, with hot licks of meadowlark song filling the air. Enveloped in the sounds and scents of the grasslands, I am a child again, holding my mother's hand, as memories of her own sage-scented prairie childhood rise up to meet her. The Great Plains grasslands are old, older than memory. For visitants like us, this ancient land offers a grounding in continuance.

Something kindles inside me when I sit on a lichen-covered boulder and realize that it has sat there for ten thousand years, ever since the retreat of the glaciers. Or when I lie on my back in the grass, soaking up the sun, and feel the Earth pressing against me as if it were holding me up. Whatever it was that lit the spark of life in the beginning of time is still present here, in the grass, wind, sunshine, and rain, in the birds and animals. Working to preserve and restore grassland ecosystems is an act of reverence for the crazy caper of life, which gave birth to us and all our flying, walking, swimming, and slithering relations. It is an expression of gratitude for the mundane gift of being here.

There is no point in pretending that everything is hunky-dory for the wildlife and wild places of the Great Plains. In my home region of the Canadian prairies alone, more than two dozen species have been added to the at-risk list since this book was first released seven years ago, and another dozen have been "uplisted" to a more critical status. Only two listed species appear to have made significant gains during the same span of years: a tiny fish called the bigmouth shiner, which has been found in new locations and is no longer thought to be at risk, and the swift fox, a cat-sized canid that went from Extirpated to merely Endangered, thanks to a long-term reintroduction effort. Meanwhile, the status of prairie birds as a group continues to worsen year by year, as formerly abundant species, like the Common (now uncommon) Nighthawk, become the focus of concern.

The fundamental problem for most prairie species is loss of habitat. To this day, we continue to lose wetlands to drainage, river-flow to dams, and both native and tame grasslands to cultivation. In the Great Plains states, for example, millions of acres of marginal cropland that were seeded to hay in the 1980s and 1990s under the Conservation Reserve Program—and that have provided living space for wildlife ever since—are currently being ploughed up for the production of biofuels. As for the surviving wild prairie, it is in declining health due to the incursions of invasive plants and the relentless, dendritic expansion of oil-and-gas exploration and other human demands. If you are looking for a place where the conservation needs are urgent and your help is required now, look no further.

. . . a strategy of "no regrets,". . . I can't promise you that a united force of grasshuggers will succeed in striking a happy balance between prairie people and the more-than-human world. It's pretty clear, however, what will happen if we do not make the attempt. From my own small experience of engagement (as board member for the Nature Conservancy of Canada and a partner in a restoration project, among other things), I can tell you that, even though the context is often disheartening, the work of conservation can be exciting, inspiring, and fun. The prairie ecosystem is battered, but it is also adaptable and tough. Repeat after me: Things can change for the better.

There is no way to hold back the future. But we can shape the course of events by engaging—fully, deeply, and passionately—with the present. And so we begin again.

Eastend, Saskatchewan
April 2010

{ ACKNOWLEDGMENTS }

THE AUTHOR IS deeply indebted to the dedicated and creative people who contributed their talents to this project, in particular photographer James R. Page, artist Joan A. Williams, editor Nancy Flight, designers George Vaitkunas and Naomi MacDougall, and cartographers Lorena Patino and Diane Perrick. This book owes much of its beauty and merit to their efforts.

The information presented here is based on scientific research conducted by biologists at universities, research institutes, and other publicly supported agencies in every province and state across the Great Plains. Although there are too many of them to mention individually, they have each earned a vote of thanks from everyone who loves the grasslands. The experts who served on the advisory panel for this project deserve special acknowledgment, since they found time in their busy schedules not only to read all or part of the manuscript but also to offer many invaluable corrections and clarifications. Many others generously shared their time and knowledge, notably: Steve Adair, Ducks Unlimited; M.G. Anderson, Brian Gray, and Rhonda McDougall, Ducks Unlimited Canada; James Bassinger and Brian Pratt, Geology Department, University of Saskatchewan; Barrie Bonsal, National Hydrology Research Centre; Jan Bednarski, Benoit Beauchamp, and Arthur Sweet, Geological Survey of Canada; Louis B. Best, Department of Natural Resource Ecology and Management, Iowa State University; Dean Biggins and Mike Lockhart, U.S. Fish and Wildlife Service; Peter Blancher, Bird Studies Canada; Don Buckle, arachnologist, Saskatoon; Bonnie Chasteen, Missouri Department of Conservation Grow Native! Program; Bob Clark and Dean Nurnberg, Canadian Wildlife Service; Meredith Cornett, Minnesota Chapter of the Nature Conservancy; Robert R. Cox, Jr., and Gary L. Krapu, U.S. Geological Survey; Sarah Davies, American Prairie Foundation; Art Davis, Cedric Gillott and Robert Randell, Department of Biology, University of Saskatchewan; Cyndi Evans and colleagues, Prairie State Park, Missouri; Gene Fortney, Nature Conservancy of Canada; Curtis Freese, World Wildlife Fund U.S.; David A. Gauthier, Canadian Plains Research Center;

Rob Gardner, Society of Grassland Naturalists; Robert Gordon, Northern Plains Entomology; Robert Graf and Owen Olfert, Agriculture and Agri-Foods Canada; Grant Harper, geologist, Thornhill; John A. Harrington, Department of Geography, Kansas State University; Kirk Henderson, Iowa Native Roadside Vegetation Center; E.H. Hogg, Canadian Forest Service, Natural Resources Canada; Sam James, Department of Biology, Maharishi University of Management; Kris Kendell, Alberta Environment; Anna Leighton, Saskatchewan Native Plant Society; Steve Malins, Banff National Park; Richard Manning, author and board member of the American Prairie Foundation; Mary Ann McLean, Department of Life Sciences, Indiana State University; Sue Michalsky, range ecologist, Eastend, Saskatchewan; Russ Miller, general manager of the Turner ranches; Wendell Morrill, Department of Entomology, Montana State University; Heather Musgrove, Saskatoon; Dave Naugle and Thomas M. Power, University of Montana; Kevin Murphy, Saskatchewan Environment; Mike Phillips, Turner Endangered Species Fund; Keith Roney and Ron Tillie, Royal Saskatchewan Museum; John Sidle, U.S. National Grasslands; James C. Trager, Shaw Nature Reserve, Missouri; Elaine Wheaton, Saskatchewan Research Council; and David and Lynn Zahrt of the Country Homestead Bed and Breakfast in the Loess Hills of Iowa. I am grateful to you all.

Finally, it is a particular pleasure to acknowledge the day-in-and-day-out contribution of Keith Bell, whose generosity and love of life have sustained this endeavor at every step.

one

WHERE IS HERE?

"You cannot take care of what you cannot see".

DR. GEROULD WILHELM, SPEAKING AT THE NORTH AMERICAN
PRAIRIE CONFERENCE, KIRKSVILLE, MISSOURI, 2002

THERE ARE PEOPLE who think of the prairie as boring, and it is hard not to pity them. We see them on the highways, trapped inside their cars, propelled by a burning desire to be somewhere else. But even as we wonder at their hurry, we have to admit that these disgruntled travelers are following in a grand old North American tradition. On both sides of the Canada–U.S. border, prairie bashing is as old as the written record. In 1803, for example, when the United States was contemplating the acquisition of the lands west of the Mississippi River from the French, through the Louisiana Purchase, the great orator Daniel Webster was moved to object. "What do we want with this vast, worthless area," he thundered, "this region of savages and wild beasts, of deserts of shifting sands and whirlwinds of dust, of cactus and prairie dogs?" And even after this supposedly howling wilderness had been annexed to the U.S., many observers remained unimpressed. The painter and naturalist John James Audubon was among them. In 1843, we find him traveling up the Missouri River on his first visit to the Great Plains. Forced onto the shore when his steamboat became grounded on a sandbar, he turned a disparaging eye toward the Dakota countryside. "The prairies around us are the most arid and dismal you can conceive of," he wrote. "In fact these prairies (so called) look more like great deserts."

Another traveler of the same era, a trader named Rufus Sage, was even more direct: "That this section of the country should ever become inhabited by civilized man except in the vicinity of large water courses, is an idea too preposterous to be entertained for a single moment." North of the border, Captain John Palliser, who crossed the Saskatchewan prairies in the late 1850s, was of much the same mind. Forget farming, he recommended. This country is just too dry.

It wasn't until near the end of the nineteenth century that the tide of expert opinion turned and the Great Plains were opened to agricultural settlement, now touted far and wide as the new Garden of Eden. The fact was, however, that these magnificent grasslands were neither desert nor garden but something completely new to European and Euro-American experience. So new that at first there wasn't even a name for them in either French or English. Pressed to come up with something, the early French fur traders had extended their term for a woodland meadow—*une prairie*—as a kind of metaphor for this big, wide, sparsely wooded, windswept world. But the Great Plains were far more than a meadow. What the travelers had encountered was a vast, dynamic ecosystem, a kind of tawny, slowly evolving organism that, in a climate of constant change, had sustained itself ever since the retreat of the glaciers thousands of years before. In the presence of this strangeness and grandeur, words and vision failed.

When the newcomers looked around them, all they could see was where they weren't. This was not forest or sea coast or mountains; it was nothing but

One of the special beauties of the prairie is the cycle of four distinct seasons, each of which remakes the landscape in its own image.

The prairies have often been described as a landscape reduced to the barest essentials of land and sky—a place where the eye is lost to distant horizons and nothing much happens. But what this depiction misses is the color and excitement of the prairies seen close-up and the rewards that come with a little knowledge and observation. As evidence, here are a few brief excerpts from my notes:

Aspen Cut, November 11: "A bright blue sky, fresh snow, sparkling and mild. We are standing at the edge of a wooded draw, looking across it to choose our route down. My companion says, 'A coyote!' and points to the opposite slope. But it's not a coyote. It's a cougar: reddish-brown, stocky, rounded head, long heavy tail, smooth, smooth movements. It flows up the slope—pauses to look back several times—then over the ridge and out of sight. On a snow-covered log at the bottom of the draw, we find large round tracks with pin-prick claw marks above the central toes."

Chimney Coulee, June 29: "Last night, we stood on a hillside, ankle deep in prairie wool, and heard a whispered quivering sound that seemed to come out of nowhere. And again, like a sudden sigh. Finally we saw them, high up over our heads, a pair of night-hawks that sometimes interrupted their insect-hunting maneuvers to plunge head-long down the sky and rasp the air with their wing feathers. In that moment, the whole place was shot with silver."

Grasslands National Park, July 9: "In just two days we have seen meadowlarks, horned larks, Sprague's pipits with their surprising pink feet, phalaropes spinning and dipping like wind-up birds in a dugout, sharp-tailed grouse, nighthawks, western and eastern kingbirds, golden eagles (nestlings and mature), jack-rabbits—huge, hallucinatory—cottontails crouching in the shade of large rocks in a prairie dog colony, mule deer, white-tails, pronghorns (bucks, cows with calves), painted lady butterflies, monarchs, showy milkweed in full bloom, prickly pear cactus with waxy yellow flowers, jumping cactus stuck to our dogs, pincushions topped with electric pink blossoms, purple prairie clover, silverleaf psoralea, brown-eyed susans, skeleton weed, blue Missouri milkvetch against bone-white clay, needle-and-thread grass, awned wheatgrass shining in the wind. The cold white glare of a full moon."

light and grass, the Big Empty in the middle of the continent. A vacant space, as they saw it, in desperate need of improvement. And this failure of vision—this inability to see and appreciate the Great Plains grasslands for what they truly are—has continued to plague our perceptions right down to the present. Flat? Boring? Lifeless? Nothing could be further from the truth. It's time to drop out of the fast lane and give the prairies, our prairies, a second, loving look.

An Empire of Grass

The key to everything that happens on the prairies lies trampled under our feet. Although grasses may look humble, they are actually versatile and tough, capable of growing under the widest possible range of conditions. Anywhere plants can grow, grasses are likely to be on the scene, whether coexisting with cactuses in a desert, poking up among lichens on the Arctic tundra, or hiding in

The word "prairie" entered the English language in the 1680s, when fur traders first began edging across the North American continent. Initially, the term was applied to the area just west of the Mississippi River, where the grasses often grew so tall that a man mounted on horseback could not see over them. Later, as the Europeans pushed farther westward, they found themselves in a country of short, spiky plants, quite different in appearance from the Mississippi grasslands. To mark this distinction, the arid grasslands of the western plains were often referred to as "steppe," a word the explorers borrowed from Russian. The term "prairie," or "true prairie," was reserved for the grasslands that the traders knew best, the tall, waving grasses of the eastern plains.

Although biologists continue to find it useful to classify grasslands by height—as short, tall, or mixed—they have dropped the old idea of true prairie. In contemporary usage, the terms "prairie" and "prairies" refer to any expanse of land that is dominated by grass and other nonwoody plants. Prairies, simply put, are grasslands. With the addition of the definite article, "the prairies" also serves as a regional designation for the great grasslands that sprawl across the interior plains of North America.

The geographical terms that are used to define the plains also require clarification. Traditionally, geographers have divided the prairie region into two components: to the west, the Great Plains and to the east, the Central Lowlands. But because there is no clear geographical feature to separate these zones, the boundary between them has never been fixed with precision. On some maps, the dividing line cuts along the 100th meridian; on others, it shifts east to follow the curves of the Missouri River. In either case, the line divides the west from the east, separating prairie from prairie. Several recent sources, however—including the online *Atlas of the Great Plains*—have erased this artificial division and redrawn the map to show the grasslands of the interior plains as a coherent unit. In this book, the term "Great Plains" refers to the grasslands at the heart of the continent, as shown by the maps on pages 6 and 22.

the leafy understory of a forest. And when circumstances are especially favorable for them—for example, when the climate strikes just the right balance between precipitation and drought—grasses can assert themselves to become the dominant vegetation. ("Dominance," in this case, refers to the plants that contribute the most living tissue, or biomass, to the ecosystem. As trees to forest, so grasses to grasslands.)

A glance at a map of the world's major grasslands suggests that these conditions are most likely to occur on a broad, landlocked plain, far from any significant body of water, somewhere near the center of a continental land mass. It is in this semiarid environment—too wet to be a desert and too dry for forest—that grasses gain the upper hand, whether it be on the steppes of central Asia, on the pampas of Argentina, on the savannas of East Africa, or in the broad heartland of North America.

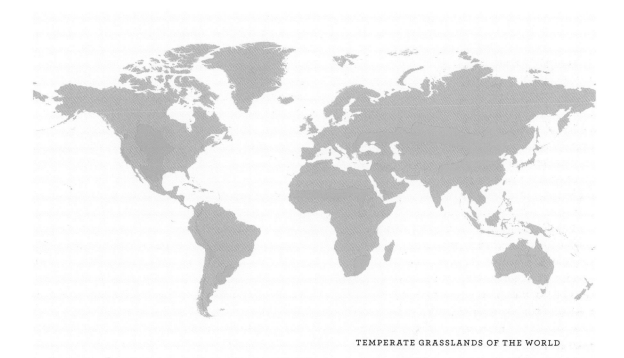

TEMPERATE GRASSLANDS OF THE WORLD

Globally, grasslands are among the largest of the Earth's terrestrial biomes, or life zones, with a sweep that covers more than a third of the land area of the planet. (At least, that's the area over which grasses would potentially hold sway if natural conditions were allowed to prevail.) We're talking some 17.8 million square miles (46 million square kilometers)—almost three times the area of Russia. In North America alone, grasslands naturally extend over about 1.4 million square miles (3.5 million square kilometers), an area larger than many of the world's major nations.

The first European known to have set foot on this great empire of grass was a soldier and sometime explorer named Francisco Vásquez de Coronado. Dispatched from Mexico City in 1540, he was supposed to investigate rumors about a kingdom called Cibola, somewhere to the north, and to plunder its Seven Cities of Gold. When these glittering mirages turned out to be sun-baked Zuni pueblos in what is now New Mexico, he turned his attention to the uncharted Great Plains, where the fish were as big as horses, the people ate off golden plates, and the king was lulled to sleep at night by a tree full of golden bells. At least that's what people told him and what he chose to believe. And so off set Coronado, with a party of armed men, in the vague direction of

Palouse Grasslands
Great Plains Grasslands
Montana Valley Grasslands
California Grasslands
Western Gulf Coastal Grasslands
Prairie-and-Oak Transition
Peace River Parklands

0   500 kilometers

TEMPERATE GRASSLANDS
AND SAVANNAS OF CANADA
AND THE UNITED STATES

present-day Kansas. In the end, the promised golden city turned out to be a village of grass-thatched huts, where the people lived by hunting bison and growing gardens, each in their season.

Yet despite this disillusionment, Coronado and his party were astonished by what they found along their route. Here lay "a wilderness in which nothing grew, except for very small plants," but which nonetheless was teeming with million upon million of strange humpbacked cattle. "I found such a quantity of cows [bison]," Coronado reported, "that it is impossible to number them, for while I was journeying through these plains, until I returned to where I first found them, there was not a day that I lost sight of them." Following

along after these apparently endless herds were parties of nomadic hunters—
ancestral Lipan Apaches, or Quechero Indians—who dressed in bison-skin
clothing (sewn with bison sinew, drawn through a bison-bone awl), slept in
bison-hide tipis, and subsisted on a diet of bison blood and bison muscle. Even
the grass in this new world was cause for amazement, as it rebounded from the
conquistadors' steps and erased the trace of their presence. In this great round
world, all that glittered was grass and an ecosystem of such richness and diversity that it could scarcely be credited.

But think how amazed Coronado would have been if he had somehow
been able to sense the true extent and variety of North America's grasslands.
Little did he know that he had set foot on a vast prairie heartland—a continent
of grass—that was flanked on every side by smaller islands of grasslands and
prairie-to-forest transitions, or savannas. To the north, for instance, beyond
his farthest imaginings, lay the Peace River Parklands, a region of rolling grass
and poplars that marked the frontier between the Great Plains grasslands
and the boreal forest. To the east, the Prairie-and-Oak Transition Zone—a
tongue of prairie interspersed with groves of hardwoods—extended to the
Great Lakes and beyond, marking the interface between the grasslands and
the eastern deciduous forest. To the south, the prairies merged and melted
into sultry, soupy marshlands to produce the semitropical vistas of the Western

Overlooked here by
the Sweet Grass Hills
of northern Montana,
the sleek little Milk
River takes the
measure of the Great
Plains, as it flows from
southern Alberta into
the Missouri River
and onward to the
Gulf of Mexico.

Gulf Coastal Grasslands. And to the west, in the broad valleys of the western Cordillera, lay the California Grasslands—spangled in spring by lupines and yellow-orange poppies—and the arid Palouse Grasslands of the Great Basin. Dominated by scraggly stands of sagebrush and spiky, sparse grasses, the Palouse, or bunchgrass, prairie stretched along the drainage of the Columbia and Snake rivers to intergrade with the shrubby growth of the Montana Valley Grasslands.

And in the center of everything there was the main attraction, the Great Plains Grasslands themselves, a landscape that even today invites wonderment. This truly is big sky country, with horizons that extend from the boreal forests of Alberta, Saskatchewan, and Manitoba to the deserts of the American Southwest and from the foothills of the Rockies to the Mississippi drainage. The numbers speak for themselves. Length: 1,500 miles (2,400 kilometers). Width: between 400 and 700 miles (between 600 and 1,100 kilometers). Vaguely triangular in outline, the region is broadest toward the north and narrows to its apex in the Hill Country of central Texas. Total area: 1 million square miles (2.6 million square kilometers), or roughly 14 percent of the entire land mass of Canada, Alaska, and the Lower Forty-Eight States.

The Grand Geographical Tour

But length and breadth are not the only descriptors of the Great Plains. The prairies also have a vertical rise and run that add a whole other dimension of interest. Formed primarily by sediments that washed out of the Rocky Mountains millions of year ago, the landscape slopes away from west to east, stepping down from an elevation of roughly 1 mile (about 1,700 meters) above sea level at the base of the foothills to a few hundred yards (or meters) on the banks of the lower Missouri River. Often, the change happens so gently that you hardly notice it. Who would have imagined, for example, that the drive across Kansas, from west to east, following in Coronado's path, would be downhill all the way and that you'd lose more than half a mile (one kilometer) in elevation while traversing that seemingly level state?

Overlain on this gently sloping plain are a surprising diversity of landforms. The geography of the Great Plains offers something for every taste, from fantastically sculpted badlands to craggy mountains to some of the flattest expanses of country anywhere on the planet. "I reached some plains so vast, that I did not find their limit anywhere I went," our old friend Coronado exclaimed in a

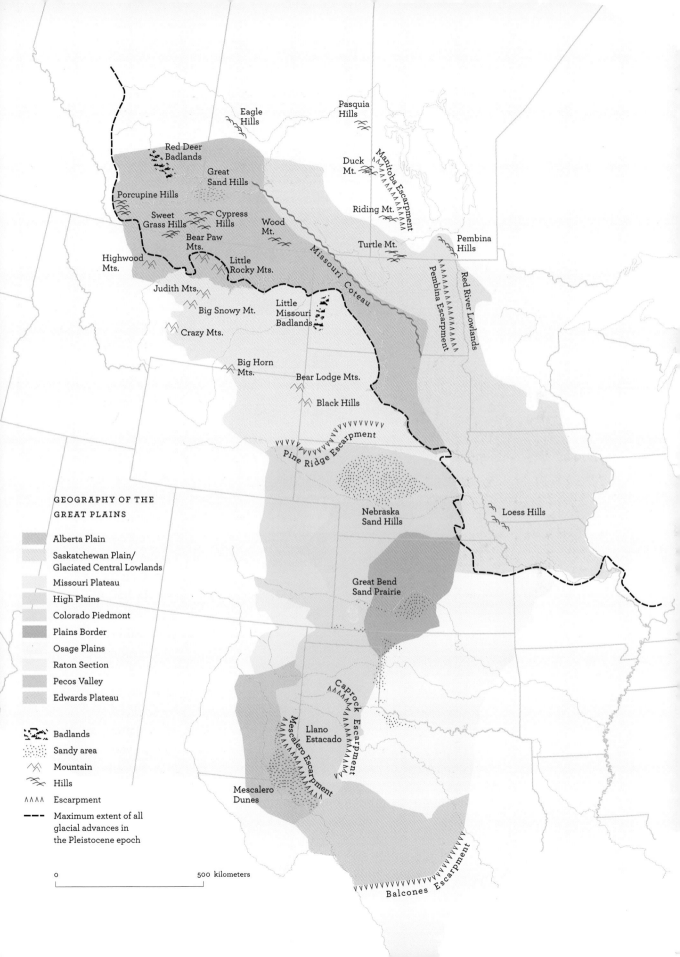

Eagle Hills

Pasquia Hills

Red Deer Badlands

Great Sand Hills

Duck Mt.

Manitoba Escarpment

Porcupine Hills

Sweet Grass Hills

Cypress Hills

Riding Mt.

Wood Mt.

Bear Paw Mts.

Little Rocky Mts.

Turtle Mt.

Pembina Hills

Highwood Mts.

Judith Mts.

Big Snowy Mt.

Little Missouri Badlands

Missouri Coteau

Pembina Escarpment

Red River Lowlands

Crazy Mts.

Big Horn Mts.

Bear Lodge Mts.

Black Hills

Pine Ridge Escarpment

Loess Hills

GEOGRAPHY OF THE GREAT PLAINS

Alberta Plain

Saskatchewan Plain/ Glaciated Central Lowlands

Missouri Plateau

High Plains

Colorado Piedmont

Plains Border

Osage Plains

Raton Section

Pecos Valley

Edwards Plateau

Nebraska Sand Hills

Great Bend Sand Prairie

Caprock Escarpment

Mescalero Escarpment

Llano Estacado

Mescalero Dunes

Badlands

Sandy area

Mountain

Hills

Escarpment

Maximum extent of all glacial advances in the Pleistocene epoch

0   500 kilometers

Balcones Escarpment

letter to the king of Spain in 1541, "with no more land marks than if we had been swallowed up by the sea.... not a stone, nor bit of rising ground, nor a tree, nor a shrub, nor anything to go by." The landscape to which he was referring is now known to geographers as the High Plains, an elevated and sometimes spectacularly featureless tableland that extends from Nebraska and Colorado into northern Oklahoma and Texas. An erosional remnant of a high-and-wide landscape that once extended over much of the Great Plains, the region is bounded on three sides by dramatic cliffs, including the upthrusting wall of the Mescalero Escarpment in the west, the tree-clad Pine Ridge Escarpment to the north, and the amazingly convoluted and striated Caprock Escarpment in the east.

To the south of the High Plains lie the limestone hills of the Edwards Plateau, or Texas Hill Country—a world in itself—where the rolling countryside is broken by domed upwellings of rock, deeply cut by streams, and eaten away underground to form a honeycomb of sinkholes and caves. The Edwards Plateau, in turn, is bounded in the south by the terraced ridges and eroded canyons of the Balcones Escarpment, which slashes across Texas at the southern limits of the Great Plains grasslands.

To the northwest of the Edwards Plateau lies the broad Pecos Valley and a landscape of spectacularly eroded caverns, sinkholes, and steep-walled limestone cuts. And north of the Pecos are the shadowed moonscapes of the Raton Section, where mesas capped with lava compete for attention with contorted badlands and the burned-out cones of Capulin Mountain and other long-extinguished volcanoes. From there, it is on to the broad, terraced valleys of the Colorado Piedmont, literally "foot of the mountains," where the waters of the Arkansas and South Platte rivers have, over millions of years, stripped away layer after layer from the original High Plains surface. (This dramatic, if localized, lowering of the surface explains, for example, why the road heading east out of Denver tracks steadily upward for the first half hour or so, as it climbs out of the South Platte floodplain and onto the surrounding High Plains benches.) The effects of water erosion can also be seen on the rugged Missouri Plateau and the deeply dissected valleys of the Plains Border region.

If water has cut into these landscapes, wind has smoothed them out. For example, the southeastern edge of the Platte River valley is softened by a broad belt of curving, undulating sand dunes that were deposited by dust storms sometime during the Ice Age. Similar formations, shaped by similar forces, are also to be found strewn up and down the drier, western side of the Great Plains, from the Great Sand Hills of southwestern Saskatchewan in the north to the

facing page: Formed by erosion sometime in the last 1 million to 2 million years, the spectacular red sandstone wall of the Caprock Escarpment forms a natural boundary between the High Plains of Texas and the rolling terrain of the Osage Plains to the east. In places, the escarpment towers as much as 1,000 feet (300 meters) above the surrounding country.

Mescalero Dunes of the Pecos Valley. And right in the middle of the map lies one of the prairies' little-known natural wonders—the Nebraska Sand Hills, a region of whale-backed, grassy rises and prairie wetlands that, at an area of 24,000 square miles (62,000 square kilometers), ranks as the largest field of sand dunes in the Western Hemisphere. These sandscapes were put in place by the relentless northwest winds that have been coursing across the landscape for millions of years.

With so few barriers to stand in their way, these same winds have had the run of the entire Great Plains region. Although their influence can be seen in many parts of the country—for example, as ridges of windblown silt along both the South Saskatchewan and upper Missouri rivers—their touch is most obvious in the eastern and southern regions of the Great Plains. These include not only areas of the Colorado Piedmont and the High Plains but also the "low plains" to the east, notably the rolling hills of the Plains Border country, the

> ### THE MOST TREMENDOUS ROARING

The Corps of Discovery led by Captains Meriwether Lewis and William Clark spent the winter of 1804–5 in the villages of the Mandans and Hidatasa, in what is now central North Dakota. It was likely there that they first heard stories about the ferocious "great white bears," or plains grizzlies, that had spilled the blood of so many warriors. But when the Corps members met their first bear, at the confluence of the Missouri and Yellowstone rivers the following spring, the kill was surprisingly easy.

"The Indians may well fear this animal equipped as they generally are with their bows and arrows," Captain Lewis mused in his journal, "but in the hand of skillful rifle men they are by no means as formidable or dangerous as they have been represented."

That was April 29, 1805. Six days later, Lewis's partner, Captain Clark, was singing a different tune. "The river rising & current Strong & in the evening we saw a Brown or Grizzly beare on a sand beech," he wrote. "I went out with one man Geo Drewyer & killed the bear, which was verry large and a turrible looking animal, which we found very hard to kill we Shot ten Balls into him before we killed him, & 5 of those Balls through his lights. This animal is the largest of the carnivorous kind I ever saw."

Plains grizzly bear

Lewis described the same encounter in greater detail: "It was the most tremendious looking anamal, and extreemly hard to kill notwithstanding he had five balls through his lungs and five others in various parts he swam more than half the distance across the river to a sandbar, & it was at least twenty minutes before he died; he did not attempt to attack, but fled and made the most tremendous roaring from the moment he was shot . . . . Capt. Clark thought he would weigh 500 lbs. for my part I think the estimate too small by 100 lbs. he measured 8 Feet 7 ½ inches from the nose to the extremety of the hind feet, 5 F. 10 ½ Ins. arround the breast."

Once widely distributed across the Great Plains, grizzlies may have been most common in the major river valleys. They were extirpated from the region by the 1890s or early 1900s.

Osage Plains, and the Glaciated Central Lowlands. Much of this sweep of country is blanketed in deep, contoured drifts of fine silt, or loess—pronounced "luss"—another gritty, wind-borne by-product of glaciation. The result is a gently undulating landscape of soft, rolling hills and, in places, extraordinary bluffs, like the delightfully eroded and unexpected Loess Hills of western Iowa.

The northern plains region, by contrast—north and east of the Missouri River, from Alberta to Manitoba and south through the Dakotas—is less apt to be buried in loess, but it nonetheless bears the imprint of the Ice Age. Here the terrain is an unmade bed of glacial rubble, or till, lying exactly where it dropped when the ice sheets retreated from the landscape ten thousand years ago. And protruding above this jumble of knobs and kettles is an assortment of sprawling, flat-topped uplands, including Turtle Mountain, Wood Mountain, and the Cypress Hills, which straddle the boundary between past and present. Like miniature versions of the High Plains, they are the last surviving remnants of an ancient, preglacial landscape that has otherwise been lost to erosion.

Finally, and most surprising of all, are the honest-to-goodness mountains that jut up out of the northern plains, particularly on the unglaciated reach of country south and east of the Missouri River. From the glowering Black Hills to the jagged Crazy Mountains, they stand as a peak experience (if you'll forgive the pun) for anyone who has been led to believe that the prairies are monotonous.

Then and Now

It is one thing to send our minds running across the contours of the Great Plains grasslands and their unexpectedly varied landforms. It is quite another to bring these spaces to life, to try to perceive them in their full, natural vitality and splendor. What would it have been like to step out onto the round bowl of the southern grasslands with Coronado in 1541, aware that at any moment our progress might be blocked by a dusty, pawing, milling herd of bison? Or, precisely 150 years later, in 1691, to have traveled with Henry Kelsey and his Cree and Assiniboine guides from Hudson Bay through the northern forest and onto the prairies of the Saskatchewan River country? What emotion would have seized us when a blocky, hunched shadow gradually resolved into the form of a massive and potentially lethal grizzly bear? Or what if we could slip back in time to 1805–6 (a mere two hundred years ago) and join Lewis and Clark on their famous expedition up the Missouri River?

Imagine: Bison beyond counting. ("I do not think I exaggerate," Lewis wrote as he crossed the Dakota plains in 1804, "when I estimate the number of Buffaloe which could be compre[hend]ed at one view to amount to 3000.") Flights of pronghorns at every turn. Elk coming up out of misty valleys to graze on the prairie at dawn. Bighorn sheep perched on the steep, crumbling walls of the Little Missouri Badlands. Wolves threading across the prairies, trailing the herds.

Two hundred years isn't very long on the geologic timescales of planet Earth. These memories lie at the very threshold of the present, so close that we half expect to be able to walk into a fold in the landscape and encounter them. And something like this still occasionally happens when we stumble across a physical trace of the past, whether it's a flaked stone tool that once belonged to a bison hunter or a shallow, saucer-shaped hollow that was worn into the dirt by generations of rolling, grunting bison. The animals have vanished, but the imprint of their flesh and blood is still on the land. It is all so mind-bogglingly recent.

There are not many places where the wild is as close at hand as it is on the Great Plains. In the Old World of Europe and Asia, no one can quite remember what "natural" looked like, because the land has been successively shaped and reshaped to meet human needs for hundreds or thousand of years. But in the New World of the prairies—right up to the moment when the settlement boom began—humans had lived off the natural productivity of this vast, sun-swept expanse of grass. From the beginning, the First Peoples had drawn their sustenance from the native animals and plants, experiencing both feast and famine as hunters and gatherers. This is not to say that they sat back passively and let nature take its course. They were active participants in the ecosystem, ready and willing to use whatever technologies they could command to improve their chances of survival. For example, they had no qualms about setting the prairies on fire, to green up the grass and draw bison in for the hunt. They tilled the soil of fertile river valleys and planted gardens of sunflowers, corn, and squash. They eagerly adapted to the new culture of firearms and horses.

Yet despite these human innovations, the underlying dynamic of the ecosystem—the interplay between climate and grasses, grazers and

> **ABUNDANCE**
> The great herd running away,
> The buffalo running,
> Their drumming hooves
> Send dust clouds billowing to the sky
> And promise good hunting
> The buffalo and her child approaching,
> Mother and Calf coming,
> Turned back from the herd,
> Promise abundance.
>
> CHAHIKSICHAHIKS (PAWNEE) SONG

predators—remained robust. A landscape that had evolved to support large herds of grazing animals was still doing exactly that, as life ebbed and flowed in time with the seasons. Then, in the early to mid-1800s, the pace of change accelerated. In far-off Washington and Ottawa, ambitious governments began to assert their claim to the land and resources of the Great Plains. As a prelude to agricultural settlement, Native people were confined on reserves and reservations, whether by persuasion or by brute force, and the bison on which they depended—the multitudes of "humpbacked cattle" that had darkened the plains—were virtually wiped out in a bloody orgy of killing. Tellingly, the final stages of this slaughter were motivated by the discovery that bison hides could be cut and sewn into leather belts and used to power machines in the burgeoning industrial complex in the East. (The last free-roaming bison were killed in Canada in 1883 and in the U.S. in 1891.) Modern times had arrived on the prairies.

And then came the settlers, an onrush of humanity that reached full flood in the late 1800s and early 1900s. Determined to make a stand in this new

The largest land animal in North America, a mature bison may stand almost 6 feet (2 meters) high at the shoulder and weigh as much as a ton. Here, a black-billed magpie takes advantage of the view, searching for insects stirred up by the bison's hooves.

Arthur Savage photo

country, the incomers quickly progressed from temporary shacks and shanties into substantial homes, making them the first people ever to establish permanent, year-round dwellings on the open plains. This was a bold experiment, occasioned with far more risk than anyone at the time seemed to recognize or, at least, was prepared to admit. But whatever the hazards, the way forward was clear. The object was to assert control over the ecosystem and redirect its natural vitality into the production of commodities that could be bought and sold on the world market. Beef, not bison. Wheat and corn instead of prairie wool.

The result of this revolution is the landscape that we see today, a colorful patchwork of fields and rangelands, where geese feed in the stubble, foxes hunt in farmyards, and meadowlarks sing their hearts out on fence posts. These are the prairies that our generation was born to, and they are beautiful in their own right.

Yet the more we love this place as it is, the more we feel the pain of what it so recently was. The wild prairie ecosystem is gone. And this tragedy is compounded by the realization that we don't even know exactly what it is that we have lost. "Civilization" and "progress" overran the grasslands with such an urgent rush that the ecosystem was disrupted before anyone had a chance to make a systematic study of exactly what was out there or to figure out how all

the pieces interacted with each other. The people who might have had the most to teach us—the last generation of hunters and gatherers—went to their graves largely unheeded by the newcomers, taking their knowledge of the prairie and its life ways with them. We are left with little to guide us except for fragments of written descriptions in the journals of explorers and early settlers—partial lists of species, brief sightings, and offhand remarks—that leave many basic questions unanswered.

The depth of our ignorance is startling. Question: How many bison were there on the plains before the slaughter began? Answer: No one can tell us with any assurance. By working and reworking the available strands of evidence, experts have estimated the precontact population at anywhere from 12 million to 125 million animals, a variance that leaves more than 100 million bison in limbo. These days, experts acknowledge that bison once numbered in the millions and probably tens of millions, but that's as far as they're prepared to go. And if we cannot account for big things like bison, how much less do we know about the smaller and less conspicuous organisms—little things like insects and spiders, fish and frogs, rodents and songbirds—that lived and died in their untold variety and interest and abundance? Yet if the wild past is lost to us, we can still look ahead. Despite everything that has happened, it is not too late to acknowledge the natural forces that continue to animate the prairie world and that, even today, shape the lives of all its creatures.

facing page: Male bison, seen here in the foreground, can be recognized by their burly physique and stout, inwardly curved horns. Females are somewhat smaller, with slender, straighter horns. What's more, in spring and early summer, they may have small red calves in tow.

Ecosystems and Ecoregions

It's not really fair to blame our ancestors for their lack of ecological awareness. At the time the plains were settled, "ecology" as we know it had not yet been invented. Instead, the science of the day was focused on fixing life to a pin, labeled and safely dead, with the species laid out in straight rows and separate compartments. (This passion for "still life" was given concrete expression in the natural-history collections of nineteenth-century museums, those great mausoleums filled with pressed plants, dried skins, stuffed birds, and mounted insects—among them, many thousands of specimens from North American grasslands.) The conception of life as a tumultuous interaction between organisms and the world around them was of no great interest to a science that was largely content to follow in the footsteps of Linnaeus, collecting things and classifying them.

But on the fringes of science, new ideas had begun to stir. In 1866, for example, an eccentric German physician-turned-philosopher named Ernst Haeckel

By late summer, big bluestem—the characteristic grass of tall grasslands—often waves above the wild flowers, like the goldenrod seen here, and reaches into the lower branches of bur oaks and other trees.

cobbled together two Greek roots—*oikos*, meaning "household," and *olgie*, or "study of"—to describe a radical new approach to life science, ecology. Drawing his inspiration from Charles Darwin's still-recent discovery that organisms are shaped by the environments in which they evolve, Haeckel described his new endeavor as the study of natural selection in action. It was, he said, "the investigation of the total relations of the animal both to its inorganic and organic environment," including its relationships, whether "friendly" or competitive, direct or indirect, with all the other organisms in its surroundings. The essence of life, he hinted, lay not in a static array of species in a collector's cabinet but in this wonderfully tangled web of interconnections.

This was a heady idea, more poetry than science, and for a long time, no one could figure out how to develop it any further. In the end, it took a pioneering biologist from Nebraska by the name of Frederic Clements to ground the new field in observation and hands-on evidence. With the native vegetation under assault all around him, Clements focused his attention on the ability of the wild prairie to recover after it had been burned over, plowed up, dried out, or otherwise disrupted. In a study published in 1916, he concluded that the prairie was a self-healing system in which, given the chance, groups of plants grew back in an orderly sequence, each wave creating the conditions required by the next, until the vegetation reached a stable configuration, or "climax." Based on his observations, Clements proposed that not only the Great Plains grasslands but the entire living world was sustained by these self-organized, internally motivated processes of renewal.

Clements's prairie-inspired theory of "community succession" made a stir in the intellectual capitals of the world. And it was there, in London, England, that his intuitions were eventually brought to a new stage of development. The concept of the "ecosystem"—the idea that the Earth operates as a series of self-organized complexes in which all components (both living and nonliving) are linked—was first advanced by British ecologist Arthur Tansley in 1935, in a self-conscious attempt to marry the study of nature with advances in math and systems theory. Suddenly, the young-and-still-somewhat-woolly-headed science of ecology, which had heretofore made do with hazy notions of "relationships" and "community," could gird itself in the vocabulary of the physical sciences. Instead of a bewildering clutter of sensations, the living world had become a "complex, adaptive, open, nonlinear system," complete with "feedbacks" and thermodynamic "fluxes."

Yet for all this new exactitude (which has served the science well), the idea of the ecosystem has proven reassuringly resistant to precise definition. As Tansley himself acknowledged, "the systems we isolate mentally are not only included as parts of larger ones, but they also overlap, interlock and interact with one another." The task of charting these overlapping and fluid realities—of acknowledging the differences between particular localities without denying their interconnectedness—remains a major preoccupation of ecologists. It is a challenge that, over the last thirty years, has inspired a continent-wide spree of "ecological land classification" and mapping. At its heart, this is an attempt to see beyond the human impacts of the last few centuries and uncover the enduring components of the environment (climate, soils, landforms, vegetation, and so on) that make one part of the continent biologically different from the next. Although we cannot go back in time and view the wild prairie in full bloom, we can attempt to identify and assess the factors that, over the long term, made them what they are or, at least, what they once were.

Ecological mapping began in the 1960s and 1970s as a relatively straightforward attempt to examine the relationship between natural vegetation and

Rough fescue

facing page: Subtle variations in soil, precipitation, drainage, and growing season determine which species of plants thrive in different regions of the Great Plains. The mixed grasslands of southern Saskatchewan, for example, are dominated by needle-and-thread grass and spangled with wild flowers like ascending milk vetch.

climate. Nowadays, however, with satellites to assist with mapping and computers to do the grunt work, the possibilities have spiraled. Instead of being limited to single variables, researchers can now consider the interplay among dozens of different ecological components all at once, factoring in everything from climate to geology, hydrology, physiography, soil chemistry, vegetation, species diversity, and where relevant, human effects on the environment. Yet even at this high pitch of sophistication, scientists have not been able to pin the living world neatly to the wall, sure that they have mapped its subtleties once and for all.

Over the last decade, the ecological regions of the Great Plains have been mapped in different ways by different agencies, whether in broad strokes as part of continentwide research or more minutely, state by state and province by province. One result of this effort is a set of maps created under the joint authorship of the World Wildlife Fund Canada and its counterpart in the United States, the WWF–U.S. Seen through the lens of these organizations, the Great Plains grasslands come into focus as a mosaic of fifteen eco-regions. The Aspen Parklands ecoregion, for example, lies across the midriff of the Prairie provinces (Alberta, Saskatchewan, and Manitoba) and provides an interface between the boreal forest and the open plains. Rising in the north as a closed poplar woodland with occasional stands of spruce, it gradually unfolds into a rolling grassland dotted with aspen groves and dominated by various spear-grasses, wheatgrasses, and most notably, fescues.

Because of the predominance of fescues in the Aspen Parklands, the region is sometimes known as a fescue grassland. The same term is also applied, for the same reason, to the community of plants found in the Foothills Grasslands. Located on the undulating slopes at the foot of the Rocky Mountains, this ecoregion is dominated by rough fescue, together with lesser quantities of June grass, speargrass, wheatgrass, and various flowers and shrubs.

East of the foothills and south of the Aspen Parklands lie two ecoregions with subtly different characteristics. The Northern Mixed Grasslands takes its name both from its northerly location and from its characteristic mixed—which is to say, mixed-height—cover of grasses. Here, the sparse, tufted vegetation of the foothills gives way to a groundcover of ankle-high grasses, notably blue grama, intermixed with an overarching canopy of knee-high stands, including various spear- and wheatgrasses. As this mixed grassland flows southward, the canopy of taller, midheight grasses gradually thins out,

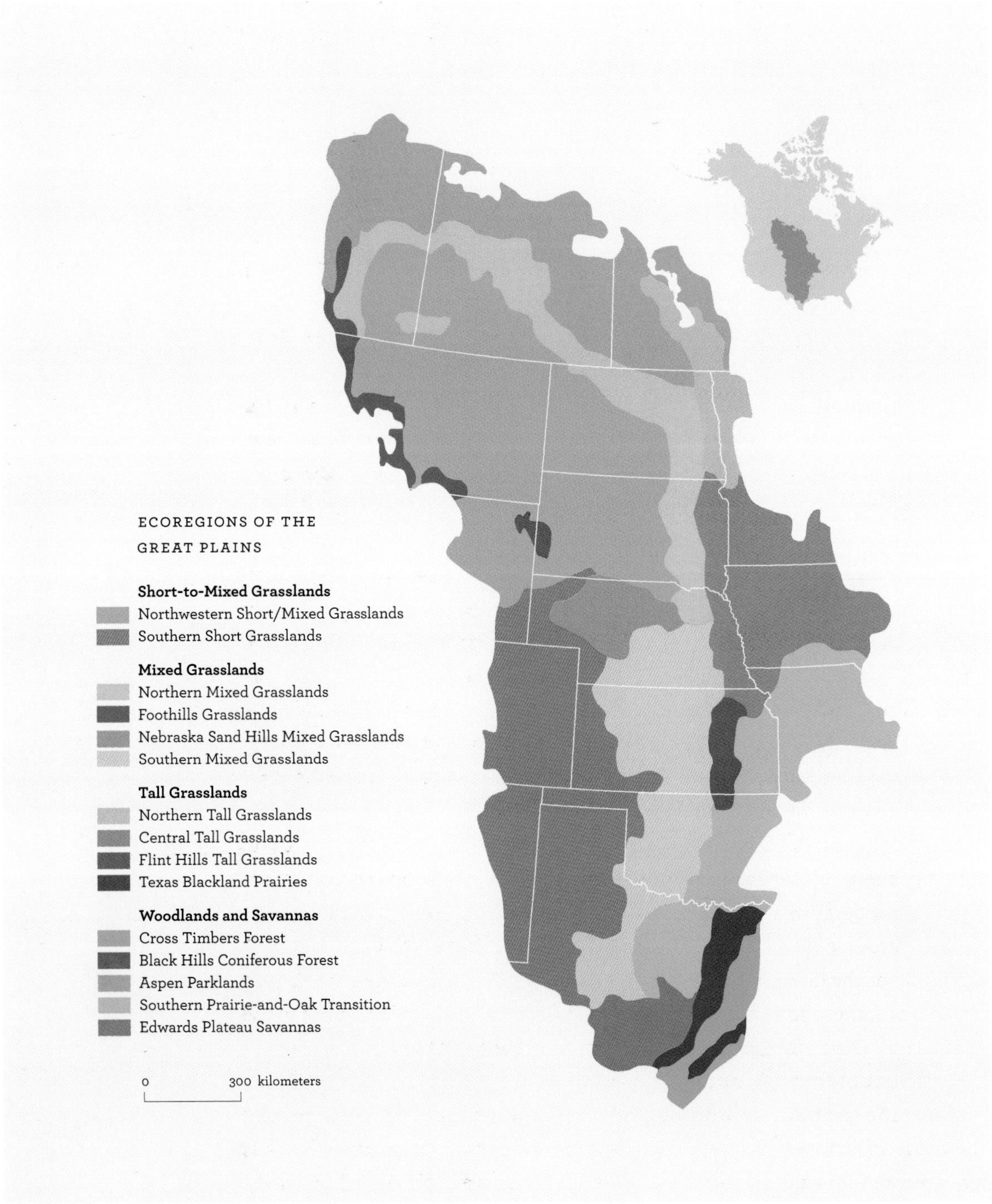

ECOREGIONS OF THE
GREAT PLAINS

Short-to-Mixed Grasslands
Northwestern Short/Mixed Grasslands
Southern Short Grasslands

Mixed Grasslands
Northern Mixed Grasslands
Foothills Grasslands
Nebraska Sand Hills Mixed Grasslands
Southern Mixed Grasslands

Tall Grasslands
Northern Tall Grasslands
Central Tall Grasslands
Flint Hills Tall Grasslands
Texas Blackland Prairies

Woodlands and Savannas
Cross Timbers Forest
Black Hills Coniferous Forest
Aspen Parklands
Southern Prairie-and-Oak Transition
Edwards Plateau Savannas

0   300 kilometers

and the overall height of the vegetation diminishes. This transition from mixed- to short-grass prairie continues to the south in the Northwestern Short/Mixed Grasslands ecoregion.

The trend to shorter grasses culminates in the Southern Short Grasslands of the High Plains. Here, in an area once memorialized by Coronado as a land of "very small plants," the vegetation is dominated by a ground-hugging mat of grama and buffalo grass. Yet here, too, there are subtle shifts. For if the grasslands diminish in height from north to south, they shoot up from west to east as they escape from the rain shadow of the Rockies. This trend is reflected in the transition from the short-grass prairies of the west, with their carpet of stunted plants, to the multilayered, knee-high vegetation of the Nebraska Sand Hills and the Southern Mixed Grasslands.

On the eastern flank of the plains lies the tall-grass-prairie region, so named for the luxuriant stands of big bluestem and Indian grass that grow—or, at least, once grew—there. "The grass is so very high that a man is lost amongst it," reported explorer Pierre François-Xavier de Chevalier as he crossed southern Wisconsin in 1761. Bright with brown-eyed susans and other flowers, these magnificent prairies extended from the Northern and Central Tall Grasslands south through the Flint Hills to the Blackland Prairies of east-central Texas. Many of the same species of grasses are also found, as an understory, in the juniper breaks of the Edwards Plateau Savannas and in the hickory-and-oak woodlands of the Cross Timbers Forest and the Southern Prairie-and-Oak Transition.

And finally, right out in the middle of everything, stand the lonely, displaced islands of ponderosa pine, white spruce, and paper birch that make up the Black Hills coniferous forest.

The ecological interactions that find expression in these varying landscapes have been at work for thousands of years. Even today, characteristics such as average temperatures, precipitation, length of growing season, and drainage patterns provide the physical framework or, one could say, the loom on which the fabric of the Great Plains ecosystem is woven. Yet for all their continuing importance, these long-term physical features are no longer the only powers in the land. Other interests have taken over; other hands are pulling threads. Those hands, of course, are human.

Over the last two hundred years, human beings have hit the prairies with the force of a major geological crisis, triggering not only extinctions and extirpations—of plains wolves, plains grizzlies, plains elk, plains bighorn sheep,

free-ranging plains bison—but also dramatic shifts in the vegetation. Taken as a whole, the Great Plains grasslands now rank as one of the most extensively altered ecosystems on Earth. There is scarcely a patch of ground where we have not left our footprints. The southernmost Short Grasslands, for instance, are as sun-baked and arid a country as you could ever expect to see, yet even there an estimated 29 percent of the ecoregion has been brought under cultivation. The surviving native prairie in the region is now devoted to livestock or converted to ranchettes, on the advancing front of urbanization. In the mixed grasslands, by contrast, the percentage of land under cultivation rises from 15 percent (in

districts with scant precipitation) to over 99 percent (where conditions are most conducive to crop production). And in the tall grasslands, with their relatively generous climate and deep, black earth, as much as 99.9 percent of the native grasses have been plowed under to make way for agriculture. Largely as a result of this destruction of natural habitat, at least 464 prairie species have declined to such rarity that their long-term survival is in question, and more names are added to the list with every passing year. (This tabulation includes only species that have been officially designated as at risk of extinction, either locally or nationally.) Of this total, a majority are unique organisms found exclusively, or almost exclusively, on the Great Plains grasslands.

These trends are deeply troubling, and we could easily get lost in the dark. To find our way forward we will have to be sure-footed, willing and able to move quickly from sorrow to hope, from past to present, from celebrating wildness to accepting and honoring our own accident-prone presence. We will need to see both the splendor of the life that has faded away and the abundance that still extends across the whole wide world of the prairie in every direction. For however diminished, the Great Plains are blooming and buzzing and wriggling and squirming with wildlife wherever we look. In the Northern Mixed Grasslands ecoregion, for example—where as much as three-quarters of the natural habitat has been lost to the plow—there are currently no fewer than 13 species of amphibians, 18 reptiles, 72 mammals, at least 160 butterflies, 222 birds, and 1,595 species of grasses, sedges, and wildflowers. This gives the region a total "species richness index," on the books of the World Wildlife Fund, of 2,095, much higher than many areas that are typically thought of as biodiversity hotspots. (By comparison, the rain forests of northern California have a richness index of only 1,710, while the Everglades come in at 1,855.)

On the Southern Short Grasslands, by comparison, where significant areas of natural grasslands remain intact as grazing land, the picture is brighter yet,

The Foothills Grasslands of Waterton National Park, Alberta, bask in the autumn sun. This ecoregion not only benefits from the trailing edge of storms borne over the mountains but also enjoys relatively mild and bright winters, thanks to the influence of chinook winds.

with 17 species of amphibians, 61 reptiles, 86 mammals, 230 butterflies, 245 birds, and an astonishing 2,359 species of grasses and other nonwoody plants, for a richness index of 3,011. Although this book can't introduce you to all those species—you'll need the appropriate local field guides for that—it will explain how this abundance of life is sustained and renewed, season after season. Far from being a sacrifice on the altar of progress that we can dismiss from our thoughts, the prairies are still very much alive and worth caring about.

About This Book

In the riotous interactions of nature, everything happens at once—sun, wind, rain, growth, birth, death—and change ripples organically through the ecosystem. For the purposes of discussion, however, it has been necessary to isolate aspects of this holistic system and discuss them one by one, each in a separate chapter. Although the subject matter is tightly interrelated, each section has been designed to stand on its own, so the chapters can be read individually and in any order. Chapter 3, "The Geography of Grass," for instance, provides a detailed look at the prairie grasses and their dynamic relationship with the extremes of a midcontinental climate. Chapter 4, "Secrets of the Soil," ventures into the dirt—a life zone all its own—and introduces a few of the strange little creatures that live beneath the ground. In Chapter 5, "Home on the Range," we come back out into the sunshine to ride through cattle country and find out how life is lived on the surviving expanses of native prairie. Chapter 6, "Water of Life," by contrast, takes us knee-deep into the nearest prairie river or pond to look into the lives of ducks, shorebirds, fish, and other aquatic organisms. Chapter 7, "Prairie Woodlands," examines the unexpectedly important role of trees in grassland ecology and asks what difference it makes that woody growth is now invading the prairies. Chapter 8, "The Nature of Farming," studies the potential and challenges of croplands as wildlife habitat. And finally, Chapter 9, "Long-Range Forecast," reconsiders the conservation status of the Great Plains—is this really the most endangered ecosystem on the continent?—and discusses a range of options for protecting and restoring its wildness.

But before we look to the future, Chapter 2, "Digging into the Past," will take us back to the very beginnings of time and the great adventure of existence.

facing page: Today's prairies are a picturesque mosaic of natural and human-altered environments, like this glowing agricultural landscape along the Kansas River.

two

DIGGING INTO
THE PAST

Time is God's way of keeping
everything from happening at once.

ANONYMOUS

THERE IS AN unseen dimension to the far-and-away spread
of the prairies, and that dimension is time. At first glance, one
might mistake this for a place that time and change have somehow overlooked. These level plains and soft, rolling hills seem
to have settled here quietly, their surface unmarred by signs of
geological strife. But appearances can be deceiving. The great
grasslands of central North America have been shaped over the
past three or four billion years by the same forces that raised
the Rockies and excavated the Grand Canyon. Their surface
has been seared by the sun, scoured by ice, blasted by blowing
sand, and buried in deep drifts of gravel. As a result of immense
energies beneath the surface of the Earth, the plains have been
raised up, forced down, drowned by oceans, and blanketed in
ash. They have experienced every shudder and wrench as continents have collided and torn away from each other, only to collide and tear away again.

The traces left on the surface of the prairies by this planetary bump and grind are surprisingly minimal. Yet if you know
what to look for and where to look for it, the subtleties of the
prairie landscape become eloquent. An oil well bears witness to
ancient tropical seas. A vast level plain provides an unexpected
reminder of the protracted violence of mountain building. A

hummocky wheat field speaks of the lumbering passage of glaciers. To an observer with a little basic geological knowledge, even the most unspectacular prairie landscape suggests a long and spectacularly interesting history.

Under the Waves

To go back into the prairie's history means to go down. The record and residue of times past lie beneath our feet, so wherever we go on the prairies, we are traveling across vanished worlds. Straight beneath you, for example, at a depth of between 2,000 and 4,000 miles (3,000 and 6,500 kilometers), lies the Earth's core—the yolk of the planetary egg—which coalesced out of a whorl of star dust some 4.5 billion years ago. This partly solid, partly fluid center is encased in an equally ancient layer of rock called the mantle. And surrounding the mantle is a covering of waxlike malleable material known as the asthenosphere, which is kept at a lethargic boil by the heat of its own radioactive decay. As the source of the molten magma that periodically shoots up through volcanic fissures and rifts in the ocean floor, the asthenosphere is the main powerhouse of geological turmoil.

The roiling-and-toiling asthenosphere occupies a zone between about 45 and 150 miles (70 and 250 kilometers) below the surface. Between it and us lies a relatively thin and fragile shell of rock, known as the lithosphere. The outermost membrane of this rocky shell is the Earth's crust, a layer that is thinner, proportionately speaking, than the skin of an apple. On the prairies, the crust extends to an average depth of 25 to 30 miles (40 to 45 kilometers). Yet this comparatively short vertical distance takes us back in time some 3.8 billion years, to an era when the flying debris of creation had begun to subside and the Earth's crust was finally able to stabilize. In this remote and inhospitable age, we find the first traces of life—microscopic stains, a few microns long, made by filaments of cyanobacteria, or blue-green algae.

Rocks from this primordial era lie right out in the open on the Precambrian Shield, but they seldom break through to the surface of the Great Plains. Instead, these ancient formations generally lie a few miles beneath our feet, providing the foundation, or "basement rock," on which the prairies have been built. Our region lies on what geologists call the North American craton, or the stable core of the continent. This is a large fragment of the Earth's crust that sheared away from an unnamed supercontinent toward the end of the Precambrian Era. By the time this happened—some 600 million or 700 million years

Trilobite

ago—the Earth (and the prairie region along with it) had already endured more than 3 billion years of mountain building, erosion, glaciation, deglaciation, and general geological *Sturm und Drang*. But things must have been starting to settle down, because when the supercontinent tore itself apart, it produced a North American continent-in-the-making that has persisted until the present.

This infant continent was not exactly the land mass that we know today. The entire western Cordillera was missing, with the result that the west coast of the craton ran south through present-day British Columbia and the Pacific states (much closer to the prairies than it is today). At first, the cratonic land mass lay exposed—a low, eroding plain, as barren as the face of Mars. But, as the geological strife continued, sea levels began to rise and the land was gradually overrun by the ocean. In time, the entire continent (with the periodic exception of a chain of tropical islands that ran diagonally across the plains, from Lake Superior toward Arizona) had disappeared beneath the waves.

For roughly the next 55 million years (from about 545 million to 490 million years ago), much of the North American craton lay under a shallow sea. Wherever the land remained exposed, it was eroded by water and wind, which ground the gritty Precambrian rocks into rounded grains of quartz sand. This sand was then swept to the coasts and out into the sea, where it settled to the bottom in beds tens to thousands of yards thick. Eventually, these lustrous sediments were overlain by layers of fine-grained mud. And whether sandy or silty, this ocean floor was literally crawling with life, particularly three-lobed, many-legged, bottom-feeding arthropods known as trilobites. After an agonizingly slow start with the cyanobacteria, evolution was finally hitting its stride, producing a menagerie of weird and wonderful undersea life. As generation upon generation of these animals lived and died, their remains settled onto the ocean

The Earth has sometimes been likened to a layer cake, in which ancient sediments are overlain by deposits from successive geological events, creating an ascending timeline from past to present.

Age	Era	Period	Epoch	Years Ago	Sample Life-Forms
	Cenozoic	Quaternary	Holocene	10,000	
			Pleistocene	1,600,000	humans
		Tertiary	Pliocene	5,000,000	mastodons
Age of Mammals			Miocene	24,000,000	rhinos, camelids
			Oligocene	34,000,000	rabbits, hares
			Eocene	55,000,000	titanotheres
			Paleocene	65,000,000	early horses
	Mesozoic	Cretaceous		145,000,000	*Tyrannosaurus rex*
Age of Reptiles		Jurassic		200,000,000	*Allosaurus*
		Triassic		250,000,000	*Stegosaurus*
	Paleozoic	Permian		300,000,000	amphibians
Age of Amphibians		Carboniferous		355,000,000	insects
Age of Fishes		Devonian		420,000,000	productids
		Silurian		440,000,000	sponges, corals
		Ordovician		490,000,000	crinoids/sea lilies
		Cambrian		545,000,000	trilobites
	Precambrian			4,500,000,000	cyanobacteria

facing page: Enchanted Rock, near Austin, Texas, was formed as a massive upwelling of molten rock during the Precambrian Era. The granite eventually solidified and now lies exposed by the eroding action of wind and water.

floor, where they were buried under thick layers of sediments. Today these fossil-rich deposits—now compressed into solid sandstone and shale—are buried some 3 miles (5 kilometers) beneath the wheat fields of the northern plains and at lesser depths in other parts of the prairies. But in a few places—like the Judith and Little Rocky mountains and the northern Black Hills—they have been pushed up to the surface, exposing their maritime history to plain view.

When the Cambrian sea finally withdrew and dry land emerged again, the forces of erosion immediately began to tear away at the newly formed rocks. But soon, geologically speaking—after a break of little more than 20 million years—the water rose and slowly spread over the land. This time, even the transcontinental island chain was bathed in the warm, clear seas. Now primitive snails munched on algae and were themselves preyed upon by giant squidlike nautiloids, with shells up to a couple of yards in length. Hundreds of new species of shelled animals evolved, including crinoids, or "sea lilies" (distantly

related to modern sea urchins), and exotic reef-forming corals. There was so much life in these oceans that when they finally withdrew some 440 million years ago, they left behind thick deposits of shell fragments and calcium-rich debris, which eventually solidified into fossil-rich limestones. These Late Ordovician deposits include the elegant Tyndall stone that is quarried in Manitoba and graces so many buildings in the Prairie provinces.

And so things continued for about the next 100 million years, as shallow oceans advanced across the North America craton, only to withdraw and then flood back in. If the run and roll of the grasslands sometimes remind us of the sea, surely this is a result of the landscape's long marine history. With every advance and retreat of the ocean, the land was burdened with fresh deposits of sand, silt, and crushed shells, which built up, year by year, in nearly horizontal, banded layers. Although some of these contrasting sea floors have since been exposed by erosion (where rivers have cut deeply down through the sediments), for the most part they lie thousands of feet beneath the grasslands.

The oceans that left these deposits behind were hospitable to life—shallow, warm, well lit, and typically tropical. During the Silurian and Devonian periods in particular (between about 440 million and 355 million years ago), these waters provided ideal conditions for reef-forming sponges and corals. In what would one day become the Canadian Prairie provinces, the reef builders of the Devonian had a heyday, constructing barrier reefs and ringlike walls that rose to heights of 300 feet (100 meters). Wherever the sea was constricted by these limestone palisades, the water gradually became super-salty. If the circulation of the sea was inhibited, water lost to evaporation could not readily be replaced, and the concentration of salts steadily increased. In time, the salts precipitated out of the sea water in these areas, leaving thick beds of potash and other minerals, notably under present-day Saskatchewan. The potash deposits in Texas were formed by a similar process but some millions of years later, during the Permian Period.

The last truly continentwide inundation withdrew from the North American craton about 300 million years ago. The next time the sea attempted to overrun the land—as it would continue to do for millions of years to come—it found itself lapping around the shores of a rocky upland that had started to rise in the eastern half of North America. Apparently, the asthenosphere had heated up and begun to force masses of molten rock up through rifts in the ocean floor. This event had sent the continental plates on a slow and perilous

Nautiloids

Crinoid/sea lily

collision course. First, Europe smashed against North America from the north-east. Then a massive supercontinent called Gondwana (the combined land masses of South America, Africa, India, Antarctica, and Australia) crunched into North America from the south, causing the land to buckle and forcing the Appalachian Mountains to lift along the east coast. The forces involved in these mighty adjustments were even felt in the middle of the craton, where a range of mountains 3,000 feet (1,000 meters) high rose out of the plains of present-day Oklahoma and Colorado. Known as the Ancient Rockies, these mountains have since been eroded to their roots by the action of water and wind.

Archimedes

Apart from the appearance of these new highlands, the west coast of the craton was comparatively untouched by these titanic collisions. Through all the commotion, the sea continued to wash up over the land, even splashing around the base of the Ancient Rockies and turning them into a cluster of south-sea islands. With every advance and retreat, the sea again left behind layers of sediment and the fossilized remains of a strange coterie of underwater life. In addition to the crinoids, corals, and other unusual beasts that had occupied Devonian waters, there were now small filter-feeders, called archimedes, that had perfect corkscrew skeletons and others, called productids, that held themselves up off the sea bottom by perching on stiltlike spines. (A wonderful jumble of 250-million-year-old sea life has been preserved in the Guadalupe Mountains of western New Mexico and Texas, which were once a complex of reefs in the western ocean.) Bony fishes swam through these waters, sometimes hotly pursued by large, saw-toothed sharks. The game of evolution was being played with feverish exuberance.

Productid

Meanwhile, back on dry land, the surface of the continent was continuing to buckle and twist. As the Appalachians were thrust upwards, land in the center

> **FOSSIL SUNBEAMS**

The modern world runs on energy that originally beamed down from the sun millions of years ago. During the eons when tropical seas lapped over the North American plains, the sun provided heat and light to sustain a thick, salty soup of algae, bacteria, and other simple forms of life. Some of these tiny creatures were capable of photosynthesis, using energy from the sun, along with carbon dioxide and hydrogen, to form glucose. When these organisms (and the others that relied on them for food) died, their energy-rich remains filtered down to the ocean floor, sometimes in and around abandoned reefs and shell middens. Here, entombed in layers of clay, they were eaten and partially digested by bacteria. Eventually, time and heat completed the transformation of solar energy into crude oil. As the black muck migrated through porous layers of rocks, it eventually found its way into reservoirs, where it collected. The pump jacks at work on the plains today are sucking up wealth that was created between 450 million and 100 million years ago.

of the craton was forced to rise along with them. A broad plain formed along the edge of the eastern highlands, sloping gently toward the western sea. When the waters receded, this coastal plain extended all the way west to present-day Alberta and Texas. And even when the sea rose up and flowed across the land, the eastern margin of the plain (roughly from present-day Manitoba south to Kansas and Missouri) was now high enough to escape all but the most severe flooding.

A new frontier for life was emerging not only in North America but on the other continents as well. Land plants, which had put in their first appearance some millions of years before, had never made much of a showing. But as stable new habitats became available, the evolutionary tree began to bud and sprout with explosive energy, producing more and larger species of land plants than ever before. In time, the soggy, boggy landscapes left by the retreating oceans were filled with riotous jungles of giant sphenopsids, or scouring rushes, tree-sized ferns, and leafy conifers. These tremendous swamps, which flourished between about 355 million and 300 million years ago, disappeared soon afterward, probably as a result of a cooling and drying trend in the climate. Buried where they fell—in modern-day Iowa, Missouri, and Kansas, among other places—the swamp plants eventually turned into coal, the characteristic rock of the Pennsylvanian, or Upper Carboniferous, Subperiod.

Through much of the next 50 million years (the Permian Period), the land shriveled in the sun. Swamps decayed, seas shrank, and the exposed plains along the west coast blew with sand and salt. But life was not to be stopped. Insects, which had dominated the wetlands of the Carboniferous, now gave rise to new dry-land forms such as beetles and the distant ancestors of crickets and grasshoppers. Amphibians, too, crawled out of the swamps and began to invent the technology they needed for life on the land—notably a soft-shelled, amniotic egg that could develop out of the water. In time, new life-forms developed that could live their whole lives on land, including massive, lizardlike creatures known as stem reptiles. Basking alongside these primitive organisms on the arid coastal plains were their near-relatives, the synapsids—the direct ancestors of modern mammals. At the root of our family tree is *Dimetrodon*, a burly, fin-backed synapsid with two stabbing canine teeth, which it used to snap up slow and unwary amphibians. We know these creatures once roamed the savannas of the western plains because wonderfully preserved skeletons of *Dimetrodon* and many of its equally bizarre contemporaries have been dug out of Permian "red bed" deposits in New Mexico, Texas, and Oklahoma.

> AGATE FOSSIL BEDS NATIONAL MONUMENT, Gering, Nebraska, features fossils of rhinos, bear-dogs, land beavers, and other animals that lived on the savannas about 20 million years ago.

> BADLANDS NATIONAL PARK, Interior, South Dakota, is rich in fossil mammals, including early rhinos, horses, pigs, and camels that date to between 23 million and 37 million years ago. A much older *Tyrannosaurus rex* skeleton, nicknamed Sue, was found nearby.

> BURGESS SHALE, YOHO NATIONAL PARK, Field, British Columbia, contains a stunning sample of the sea life that would have flourished on the plains (then the continental shelf) just over half a billion years ago.

> DINOSAUR PROVINCIAL PARK WORLD HERITAGE SITE, Patricia, Alberta, provides access to a stretch of the Red Deer Valley from which the skeletons of more than three hundred Cretaceous dinosaurs have been recovered. Many of them are housed in the Royal Tyrrell Museum of Paleontology in Drumheller, Alberta.

> DINOSAUR NATIONAL MONUMENT, near Dinosaur, Colorado, and Jensen, Utah, preserves the remains of Jurassic dinosaurs—the original Jurassic Park.

> GUADALUPE MOUNTAINS NATIONAL PARK, Salt Flat, Texas, features a 250-million-year-old reef that once lay on the bottom of the sea and now towers over the deserts and plains of west Texas.

> HOT SPRINGS MAMMOTH SITE, Hot Springs, South Dakota, contains the fossilized remains of at least fifty-two mammoths and other Ice Age creatures.

> LUBBOCK LAKE LANDMARK, Lubbock, Texas, pre-serves evidence of human activities (including hunt-ing) over the last 11,500 years.

> PICKETWIRE CANYONLANDS DINOSAUR TRACKSITE, Commanche National Grassland, near La Junta, Colorado, still bears the imprint of dino-saurs that plodded through the mud 150 million years ago.

> T-REX DISCOVERY CENTER, Eastend, Saskatchewan, focuses on one of the most complete *Tyrannosaurus* fossils ever uncovered.

> WYOMING DINOSAUR CENTER, Thermopolis, Wyoming, is devoted to the study of Jurassic dino-saurs, such as *Allosaurus* and *Diplodocus*.

Beaked synapsid

Dimetrodon

So it is that we find ourselves near the end of the Permian, watching a lump-ish, beaked synapsid called *Kannemeyeria* breaking off the tough stem of a broad-leafed conifer somewhere along the west coast of Texas. Under our feet lie the accumulated sediments of 3.5 billion years, or more than 90 percent of the geological timeline. Yet except for the wide spread of the horizon, there is little in this scene to put us in mind of the modern prairies. No grass, no gophers, no pronghorns, no playas or sloughs. Something radical will have to happen to create the landscape that we see around us today. Something revolutionary.

Terrible Lizards

The Permian Period ended in a biological catastrophe—the most severe mass extinction in all of geological history. During a period of several million years, over 95 percent of all the species living in the oceans were eliminated, together with 75 percent of terrestrial vertebrates. Why did this disaster occur? No one knows for sure, but the continuing gyrations of the continental plates may have been partly to blame. By the Permian Period, the continents had become temporarily fused into one gigantic land mass, called Pangaea. At the same time, the floor of the oceans apparently warped downward, drawing the sea away from the land and exposing a vast and inhospitable heartland of hot, dry silt and sand. These deserts had little to offer to life-forms that, in ages past, had flourished in a watery world of lagoons and swamps. Perhaps this change in conditions is enough to explain the huge loss of life. But whatever the probable causes (and many have been invoked), the impact was severe, and, despite the nonstop creativity of evolution, it would take millions of years for the Earth to repopulate itself with a full range of plants and animals.

At the same time that this biological revolution was occurring, a major geological upheaval was also underway. The continents, after docking together in Pangaea for some millions of years, began to tear away from one another. As Europe sheared off to the east and the Atlantic Ocean opened up, the North American craton was shoved slowly westward. Eventually, about 165 million years ago, the drifting continent ran into a small fragment of the Earth's crust (perhaps an island chain), known to geologists as a terrane. As the continent plowed onward, it contacted other, similar obstacles in its path. One by one, these terranes were crushed against the west coast of the craton and added to its mass. The impact of these collisions—which would continue sporadically for about the next 100 million years—caused the western margin of the

craton to fold, twist, crack, and rise up mightily, until ranges of ragged peaks ran along the length of the continent. The present-day plains (which for so long had lain along the west coast, exposed to the run of the sea) were now guarded by the serried ranks of the Rockies.

Early in this process, before the wall of mountains was complete, the ocean still sometimes slipped through gaps in the palisade and washed across the plains. This happened several times during the Triassic and Jurassic periods (between 250 million and 145 million years ago), culminating in a huge incursion, known as the Sundance Sea, which swept as far east as present-day Saskatchewan, Nebraska, and Texas. But these waters were soon expelled from much of their floodplain by a deluge of a completely different sort—an influx of mud and sand that washed down off the slopes of the newly formed mountains. No sooner had the mountains raised their heads than erosion began to level them. Mixed with generous quantities of volcanic ash from the tumult of mountain building, these sediments were strewn across the plains as far east as the Dakotas. Today they form brightly banded sandstones and shales—the Success, Kootenay, and Morrison formations by name—that bear witness to an awesome struggle among rivers, mountains, and seas. They also contain evidence of an awe-inspiring bestiary of ancient life.

The earliest amphibians appeared about 345 million years ago, as long-bodied, short-limbed animals that resembled crocodiles with finned tails. Modern frogs and toads put in an appearance somewhat later, alongside the dinosaurs of the Jurassic Period.

The cataclysmic extinctions at the end of the Permian had left a biological void, but by the Middle Triassic (about 225 million years ago), this vacuum had been filled to bursting with reptiles. Creeping, crawling, swimming, flying, stomping across the land, reptiles had become the dominant animal group on Earth. Chief among them were the dinosaurs, including the 80- to 100-ton *Brachiosaurus*, which raised its ultralong neck to browse in the treetops, and the plated *Stegosaurus*, which had seventeen trapezoidal shields of bone embedded along its spine. Unfortunately for them (but fortunately for succeeding generations of dino-enthusiasts), hundreds of these large-bodied, small-brained animals apparently tramped into the rushing rivers, got stuck in the mud, and died. Their bones were then swept away by the current and dropped on snags and in backwaters, where they lay in thick beds. These Morrison deposits provide the focus for the Dinosaur National Monuments in Colorado and Utah. Although now in the mountains, the deposits were laid down on the plains, and the same or similar species must have lounged under ginkgo trees and trudged through the spiky underbrush of what is now the Great Plains.

The sea, which in the Late Jurassic had been driven off the continent by sediment from the mountains to the west, managed to creep in one last time during the Cretaceous Period (between 145 million and 65 million years ago). By this time, the Rocky Mountains formed an unbroken dyke along the west coast, but the rising waters rushed around it at both ends, flowing south from the western Arctic and north from the Gulf of Mexico. By the time the waters met in Colorado about 100 million years ago, almost the entire prairie region was inundated. It would take another 40 million years or so for the sea to make its final retreat, but when the water was finally gone, a vast plain lay exposed, stretching farther than the eye could see across the interior of the continent.

The first eyes to gaze across those broad, unfettered vistas were no doubt reptilian. By the Late Cretaceous, herds of heavyset *Triceratops*, with their wide, frilled collars and clustered horns, were roaming across the countryside and foraging in lush stands of horsetails, ferns, gingkoes, and palms, keeping an eye peeled for their most dangerous enemy, *Tyrannosaurus rex*. In case we were in any doubt about what *T. rex* preyed upon, a paleontologist in Saskatchewan has recently found a large sample of fossilized tyrannosaur dung. The .5-gallon (2.3-liter) lump contains what appears to be the crushed head frill of a juvenile *Triceratops*. Much of what we know about Cretaceous dinosaurs, both trivial and profound, comes from sites on the Great Plains, including the Red Deer River Valley in Alberta and the Frenchman Valley in Saskatchewan.

Brachiosaurus

Stegosaurus

facing page: At home in the sagebrush country and short-to-mixed grasslands of the northern and central Great Plains, the greater short-horned lizard is descended from reptiles that lived during the dinosaurian era. It is a member of the iguana family and subsists largely on ants.

> THE THIN WHITE LINE

Did the catastrophe that killed the dinosaurs come flaming out of the skies? For the last thirty years, geologists have been mesmerized by the possibility that a huge meteorite collided with the Earth about the time the last of the great reptiles disappeared. The evidence suggests that some 65 million years ago, a hunk of rock 6 miles (10 kilometers) in diameter crashed into the ocean off the present-day Yucatan coast with the force of a 100-million-megaton bomb, creating a worldwide holocaust of toxic vapor and soot. Beads of sizzling hot glass were ejected from the blast, possibly setting the rain forests and swamps alight as far north as Saskatchewan and Alberta. Swirling clouds of ash and dust blotted out the sun, as the world settled into the gloom of "impact winter."

Yet despite the magnitude of this disaster, time has removed almost every trace. Today—apart from a 100-mile-wide (180-kilometer-wide) crater off the Mexican coast—little remains except for a narrow band of whitish clay that is visible in several dozen places around the world. Not only does this layer contain residues of the rain of molten glass, it is also distinguished by the presence of "shocked quartz" (grains of sand that have been distorted by an impact) and by a high concentration of iridium, an element that is rare on Earth but common in meteorites and other space objects.

Once derided as sci-fi fantasy, the idea that a giant meteorite struck the Earth is now generally accepted by experts. But is that really how the large dinosaurs met their end? New findings suggest that the answer may be "yes." Analysis of fossilized pollen from in and around the impact zone appear to reflect a sudden die-off of photosynthetic plants. Perhaps, in the dark days after the collision, food webs collapsed and great beasts like *T-Rex* and *Triceratops* starved to death. Key sites for answering these intriguing questions include the Cretaceous/Tertiary, or K/T, Boundary strata in the Red Deer Valley at Drumheller, Alberta; the Frenchman Valley in southwestern Saskatchewan; Dogie Creek in eastern Montana; and Badlands National Park in South Dakota.

Triceratops

Tyrannosaurus rex

Succored by a mild and equable climate, much improved since Permian times, life in the Cretaceous was full. The air thrummed with insects, including moths and bees. The massive flying reptile *Quetzalcoatlus* rode the updrafts over the southern plains on a span of 35- or 40-foot (11- or 12-meter) wings—wider than those of a single-engine plane—searching for the bodies of the dead and dying. Frogs and salamanders hid in the underbrush, a habitat they shared with cowering, timorous mammals, few of them bigger than mice. Yet by the end of the Cretaceous Period, about half of this rich assembly of species—including all of the flying reptiles and the dinosaurs—had completely disappeared. And again, the reasons for this mass extinction are unclear. Most geologists believe that the Earth was hit by a huge meteorite that obscured the sun with a thick cloud of dust. Others argue that a sequence of volcanic explosions in India and elsewhere had a calamitous cooling effect on the climate. Still others believe that a gradual deterioration of the climate, over several million years, eventually put paid to the terrible lizards. But whatever the cause or causes, the outcome was clear. The dinosaurs were gone, and the plains were ready and waiting for a new group of pioneers to make themselves at home on their wide open spaces.

High and Dry

Not long after the last dinosaur drew its final breath, something strange began to happen along the western margin of the Great Plains, in the heart of present-day Montana and Wyoming. About 50 million years ago, for reasons that no one can explain (more crashing and grinding off the west coast?), the level plains of the Cretaceous seabed began to heave upward, bend, and in places, crack open. Molten rock from the asthenosphere bubbled up through the fissures, sometimes crystallizing before it reached the surface, sometimes pouring out across the land to form dykes, domes, and ridges of lava. When the smoke cleared, mountains stood right out in the middle of the level plains. Subsequently honed by erosion, these unexpected rocky peaks still punctuate the western landscape from the Sweet Grass Hills east through the Bear Paws and the Little Rocky Mountains, and south to the Crazy Mountains and the Black Hills.

At the time of their formation, the isolated "prairie mountains" did not have the presence that they do today. Even the main ranges of the Rockies were little more than bumps that protruded above a muddy, gravel-strewn landscape. The higher the mountains had thrust themselves up, the faster erosion had worn

› PRAIRIE MOUNTAINS

The Black Hills, which straddle the border between Wyoming and South Dakota, are the most easterly outliers of the Rocky Mountains. (On some geological maps, they are actually identified as the Central Rocky Mountains.) These hills rose up out of the plains between 62 million and 48 million years ago. The core of the hills, extending roughly from Deadwood to Wind Cave, is a massive, elliptical dome of granite, some of it as much 2.5 billion years old, that was thrust up from deep underground and exposed by erosion. Now carved into spires and peaks, the dome lies shoulder deep in a broad, encircling apron of younger rocks, formerly ocean floors, that is known as the Limestone Plateau.

At one time, about 37 million years ago, this plateau was completely buried in sediments that had washed down off the dome, but that overburden—and much more besides—has since been washed away. Erosion has also created the remarkable Racetrack, or Red Valley, that runs around the foot of the plateau. But nothing has been able to wear away the hogback of resistant Cretaceous sandstone that encloses both valley and hills like

Deadwood

Sandstone
Hogback

Granite Core
Area

Mount
Rushmore

Limestone
Plateau

Wind Cave
National Park

Red Valley

a fortress wall. Chosen by Americans as a site to honor past presidents (four of whom are represented on Mt. Rushmore), the Black Hills also stand as a natural monument to the colossal energies that shaped the continent.

them away, until they lay buried, neck deep, in their own shed silt, sand, rock, and clay. (The thick coal deposits in the Powder River Basin of northeastern Wyoming were formed when tons of this muck overran a peat bog some 50 million years ago and buried the vegetation under 10,000 feet, or 3,000 meters, of sediment.) Year after year, rivers carried a massive tonnage of this debris eastward to the central plains, depositing it onto a broad, eastward-sloping alluvial fan. As the braided streams of the floodplain washed over the sediments, they gradually licked the surface smooth, creating a landscape that in places is so level that it almost seems supernatural. This stunning flatland once extended from the knob-peaked Rockies across southern Alberta and Saskatchewan, south through the eastern Dakotas, east to the Flint Hills of Kansas, and down to central Texas. Today, though much diminished by erosion, this landscape

facing page: The mysterious boulders, or concretions, at Red Rock Coulee near Medicine Hat, Alberta, were formed on the floor of a shallow, inland sea about 75 million years ago, during the Cretaceous Period.

persists as the High Plains of Nebraska, Kansas, Oklahoma, and Texas. Legend has it that when the Spanish first crossed the plains of Texas in the 1500s, they used stakes to mark their route because the land was so spectacularly feature-less. Hence the name Llano Estacado, or the Staked Plains, of northern Texas.

Some 45 million years ago, when the High Plains landscape was still being shaped, it would have taken more than stakes to help travelers find their way, for it was covered by a dripping, tangled forest. Globally, the climate had never been more amenable to life—there were dawn redwoods near the North Pole—and the plains basked in warm, wet, subtropical weather. A lush wood-land spread across the midcontinent, alive with an impressive variety of birds

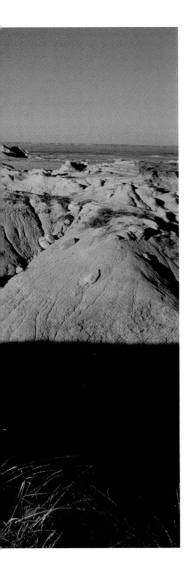

and mammals. Ancestral squirrels and monkeys leaped through the overstory, while down below, titanotheres—beasts the size of rhinos, with knobby horns and sharp tusks—shuffled across the forest floor feeding on shrubs. Among the other browsing animals of the time was an early ancestor of the horse, *Orohippus* by name, which had four toes on its front feet and three on the back and grew to be about the size of a large Shetland sheepdog.

Life was easy. But then a sequence of unrelated events halfway around the world sent the climate into a nosedive. (According to one theory, the separation of Antarctica from Australia caused a major rerouting of oceanic currents, with the result that water from the poles no longer mingled with water from the equator. The South Pole thus became an isolated refrigeration cell that eventually spread a chill around the entire planet.) Beginning about 37 million years ago, the average global temperature dropped by 14°F (8°C) over the span of a million years. Thereafter, despite brief periods of recovery, the climate continued to cool. As the weather became cooler and drier, the tropical forests of the North American plains began to wither and die away.

But conditions that were death for palm trees were ideal for another group of plants. Relative newcomers on the evolutionary scene, grasses had first appeared shortly after the extinction of the dinosaurs but had met with limited success. They were drought specialists, and while humid conditions prevailed, they had been confined to small patches of ground that had somehow been deprived of abundant rainfall. Now, not only were the tropical rains failing because of a global drying trend, but the North American plains were under a special disadvantage. With the Rockies in place, storms that rolled in from the Pacific tended to drop their precipitation as they swept up the western slopes. By the time they reached the plains, they were pretty much wrung out. But

Eocene herbivores

At Toadstool Geologic Park, near Chadron, Nebraska, sediments that originally eroded off the young Rocky Mountains were subsequently thrust up by faulting and then sculpted by erosion. The park also features a trackway of fossil footprints left millions of years ago by giant pigs, rhinos, camels, and other prehistoric beasts.

grasses don't require much moisture, and this characteristic gave them a competitive edge. Over the next several million years (between about 24 million and 3 million years ago), grasses gradually became the dominant plants across the Great Plains.

If we could slip through a crack in time and go back to the plains of Nebraska some 20 million years ago, we would find ourselves in a landscape that is at once familiar and wonderfully strange. This is big-sky country, an open landscape of shoulder-high grasses dotted with walnuts and other broad-leafed trees, vaguely reminiscent of the savannas of East Africa today. A broad river courses across the plain, its margins fringed by willows and its current murky with sediment from the constantly eroding Rockies. Whenever this river floods, it coats the land with yet another layer of silt and sand.

The river is the main source of water in this increasingly arid land, and wildlife flocks to its banks. Herds of miniature rhinos (about the size of domestic

One of the distinctive species of the northern and central Great Plains, the white-tailed jackrabbit traces its ancestry back to the Miocene Period, some 37 million years ago.

pigs but with two horns sprouting from the ends of their snouts) plunge into the shallows to find refuge from biting flies. Ancestral horses called *Parahippus*, somewhat bigger than *Orohippus* but still the size of dogs, come down to the river to drink at dawn and dusk. The rest of the time, they range across the savanna, plucking leaves off the trees and grazing on grasses that tower over their heads. Because grass is very abrasive, *Parahippus* have acquired specially ridged teeth that are able to withstand the daily grinding. Llamalike camels (members of a family that evolved in North America and only later migrated to South America and Eurasia) lounge in the willows but keep an eye out for any suspicious shadows moving through the bushes. In this world, danger takes the forms of saber-toothed cats and long-jawed dogs, some of them as large as coyotes and wolves. Smaller dogs, the size of foxes, prey on the *Paleolagus*, or "ancient rabbits," that burrow into the roots of shade trees, and on *Paleocastor*, or "ancient beavers," that, amazing as it seems, occupy deep, corkscrew burrows in the middle of the dry prairie.

Dwarf rhinoceros

Traces of these animals, and others like them, have been preserved at the Agate Fossil Beds National Monument on the Niobrara River in northwestern Nebraska. Here, the buried beds of bone testify not only to remarkable lives but also to miserable deaths. It seems that the drying trend, which had driven back the rain forest and allowed the lush parklands to spread, occasionally became so severe that it stressed even the savannas, causing rivers to dry up and trees to blacken. Animals gathered alongside the dying rivers and died along with them. Later, when floods flashed down out of the mountains, the currents gathered up the bones, massing them into backwaters and oxbows.

Orohippus and Parahippus against the silhouette of the modern horse, Equus

As the centuries ticked by, the climate became progressively more arid. Soon, in place of the lush savannas, a tawny, almost-treeless grassland sprawled across the plains. And although many mammalian species survived—including rhinos, horses, camels, rodents, cats, and dogs—all were challenged by their changed and unforgiving environment. An unremitting diet of grass pushed grazing animals to develop high-crowned teeth, which grew in to replace themselves as they were worn away. The absence of hiding places put a premium on speed, forcing both predator and prey to adopt the runner's long-legged physique. Hunter and hunted also came to rely on their quick wits, as the brain power of both players was augmented.

Paleocastor

As it turned out, these hard-won adaptations would offer little protection against the trauma that was about to unfold—the Ice Age.

Permanent Winter

Nobody knows for sure why the cold settled in as it did. Perhaps (as one theory suggests) the chill from the Antarctic refrigeration cell crept gradually north until the Arctic Ocean froze and exerted its own cooling force. Or maybe the expansion of the continental land mass over several hundred million years had caused the global temperature to trend downward. (Since land holds less heat than water, a larger land mass might logically translate into lower temperatures.) The rise of the Rockies and other mountain ranges around the world may also have contributed to the decline by disrupting the jet stream and causing cold Arctic air to spill south across the land. Or perhaps all these Earthbound events were irrelevant in the grandeur of space, where a wobble in the Earth's orbit and the inconstancy of the sun may have triggered subtle changes in the climatic system.

facing page: Glacial erratics—boulders of granite from the Canadian Shield that were carried south by advancing ice—still stand where they were dropped at the end of the last glaciation.

Be that as it may, between about 3 million and 2 million years ago, the Earth had cooled so much that permanent winter had settled over the northern reaches of the continent. The tepid summers no longer melted away the preceding winters' snows. Beginning at high latitudes and progressing southward, drifts built up into mounds, and mounds into mountains, until the snow compacted into ice under its own tremendous mass. Eventually, after several thousand years, these glaciers began to advance, flowing almost imperceptibly but relentlessly south over the Central Lowlands. In time, the northern third of North America was buried under some 2 miles (3 kilometers) of ice; that's about the height, from base to peak, of Mount Everest. In its heartland on the Precambrian Shield, the ice reached a maximum depth of about 16,000 feet, or 5,000 meters.

Geologists used to believe that the glaciers advanced and retreated four times over a span of about 2 million years. These successive incursions were known in North America as the Nebraskan, Kansan, Illinoian, and Wisconsin glaciations, in honor of their southernmost extent. But more recent research suggests that the glaciers probably made many more than four sweeps down the continent, each time grinding away the traces left by previous glaciations. Since much of the record has been wiped clear, a detailed chronology of the Ice Age on the prairies cannot be reconstructed. But we do know that by about 1.2 million years ago, a vast slab of ice had bulldozed its way almost to the present-day confluence of the Missouri and Mississippi rivers. At its maximum, the ice sheet probably extended beyond the Canadian provinces, sweeping across

northeastern Montana and south through the Dakotas to northeastern Kansas. From there it cut across the plains of northern Missouri and then eastward, across the continent, to the ice-stricken valley of the St. Lawrence.

After that ice sheet (the Kansan) retreated, the glaciers never again penetrated quite so deeply into the plains. The final glaciation, for example, which began some 100,000 to 75,000 years ago, didn't progress much farther south than central Iowa. But the devastation that the glaciers inflicted was not limited to their actual footprint. Whenever the glaciers melted back, they left behind outwash plains of sand and silt. Ferocious winds that developed over the ice fields picked up this grit and hurled it around the interior of the continent. In a number of places (notably, the Great Sand Hills of Saskatchewan and the Sand Hills of western Nebraska) the wind laid down its burden in vast fields of dunes. Elsewhere, the storms whipped up clouds of dust—rock that had been ground into flour by the glaciers—and broadcast it over the land. Today, these silt, or loess, deposits, often several yards thick, form the bluffs along the Iowa side of the Missouri River and provide the matrix for rich, rolling farmlands in Nebraska, Kansas, Iowa, Missouri, and elsewhere.

The ice began its final, halting retreat about 18,000 years ago, a transition that marked the beginning of the present interglacial period. (Unless the buildup of greenhouse gases warms the climate enough to break the Ice Age cycle of retreat and advance, a new glaciation can be expected to begin within the next several thousand years: some experts expect it within decades.) Over the succeeding 10,000 years, a block of ice larger than present-day Antarctica gradually melted away, and it didn't go quietly. Torrents gushed from the eroding ice sheets, gouging out meandering coulees and wide flat-bottomed river valleys as they coursed eastward over the plains. Today, dry coulees writhe incongruously across the northern prairies, from nowhere to nowhere, and glacial spillways seem ludicrously oversized for the quiet streams, like the Milk River, that now occupy their broad channels.

And it wasn't only moving water that left its mark on the land. In many places, meltwater was prevented from flowing away by ice dams, and the silt-laden water pooled to form shallow, milky lakes, such as Glacial Lake Regina in south-central Saskatchewan and Lake Dakota in east-central South Dakota. The largest of these "proglacial" lakes, Glacial Lake Agassiz, flooded some 135,000 square miles (350,000 square kilometers) at its maximum extent (three times the size of Lake Superior, the largest modern freshwater lake), including extensive tracts in Saskatchewan and Manitoba and the Red River

Lowland in eastern North Dakota and northern Minnesota. When the ice and then the water finally retreated from the land, these lake bottoms stood exposed as wide flat plains bounded by terraced beaches, all covered with a dressing of mineral-rich silt.

The land that emerged directly from under the ice sheets, by contrast, was a rough-and-tumble mess, strewn with the rubble that the glaciers had dropped as they retreated. Sinuous ridges of gravel and silt, called eskers, marked the courses of streams that had once flowed under or through the ice; strange conical hills called kames stood where streams pouring out of the glaciers had deposited gravel and sand. One of the most prominent glacial features on the northern plains was a long, broken ridge of hill country, called the Missouri Coteau, that meandered (and still meanders) across central Saskatchewan and south through the Dakotas. Geologists refer to the Coteau as "dead ice moraine," because it formed when hunks of ice became buried in gravel and lay there for centuries, ever so gradually rotting away. As each block of ice melted, the gravel that had been lying on top of it sagged to form a depression, or prairie pothole.

Meanwhile, south of the reach of the glaciers, on the foreshore of the Rockies, the landscape had also been undergoing renovations. Sometime before the

> ### THE CREATION OF BEAR BUTTE, MONTANA

This a northern Cheyenne story, as told by Jessie Beaverheart American Horse of Lame Deer, Montana. Once there was a very beautiful young woman, whose father was determined to choose the right husband for her. Many men came to ask for her, but her father turned them all away. Finally, a tall, handsome young man, with two fine bay horses, appeared. "I am Cheyenne," he proclaimed. "I have been a Cheyenne as long as there have been Cheyennes." Although the father found him mysterious, he presented the young man to his daughter, who happily agreed to marry him.

After a while, the new husband told his wife, "There is one thing you must never do. Never turn your back on me." He refused to explain this rule except to say that something terrible would happen if she ever broke it.

A year passed and the couple had a son. When the child was just a few months old, his father started teaching him how to ride. The mother protested—the baby was too young—and when her husband refused to listen, she snatched the infant and ran. "I told you never to turn your back on me!" the man howled, and he took off after them. When the woman looked back, she discovered that she was being pursued not by a man but by a grizzly bear.

Desperate, she ran to the top of a little mound and called out to Maheo, Above Person, for help. Right away, the hill began to lift up into the air, carrying the woman and child to safety. By the time the bear reached the hill, it was too steep and slippery for him to climb. Today, you can still see the marks left by his claws as he tried to scramble up the butte and, if the light is right, you can even make out the moccasin tracks of the woman and little boy. It is one of the places in the old Cheyenne country where women can look for power.

Saber-toothed cat

Ice Age set in, the entire western prairies had inexplicably begun to rise. As a result, the rivers, which previously had been building up the plains with loads of gravel and silt, now began to cut through the very layers they had previously deposited. This erosion was most dramatic along the slopes of the Rockies, where the rivers were powerful enough to wear through 70 million years of sediments. Along the Front Range of the mountains in Colorado, for example, the South Platte and Arkansas rivers have dug down about 1,600 to 2,000 feet (500 to 600 meters) below the level of the High Plains, right down to—and into—the seabeds of the Cretaceous. Only where erosion-resistant layers of rock have stood against this assault can the remnants of the older landscape be seen. The tops of the buttes and mesas that tower over the eroded plains were once a part of this continuous high plain.

By the end of the glaciation, the Great Plains of North America had been transformed from the seabed of ancient times into a mosaic of distinctive landscapes. To the north extended a rumpled terrain of glacial debris. Beyond the limit of the glaciers, to the south and east, lay a softer landscape of ancient ocean floors, much of it now blanketed in wind-shaped drifts of glacial sand and silt. Across these rolling hills to the west, the flatlands of the High Plains stepped up steadily toward the front ranks of the Rockies. And everywhere, rivers were cutting down into the land, etching deep valleys, canyons and, where the land was suitably dry and bare, badlands.

But if the varied landforms of the plains were beginning to look more like those of the present, many of the life-forms still did not. Disadvantaged by the cool, wet weather of the Ice Age, the grasses that had previously dominated the plains had lost ground to other plants. Now a band of tundra skirted the retreating ice, while to the south, dark coniferous forests spanned much of the continent. Pure grasslands were restricted to scattered meadows and, perhaps, to a relict prairie crammed into the southernmost plains. Together, these diverse habitats were occupied by a stunning array of life, including white-tailed and mule deer, caribou, several species of pronghorns, black bears, cougars, bobcats, lions, cheetahs, saber-toothed cats, horses, llamas, one-humped camels—even Ice Age elephants. Woolly mammoths (shaggy beasts that stood 10 feet, or almost 3 meters tall) browsed on the tundra, while Columbian mammoths (just as unkempt and much larger) appear to have favored the remnant patches of grassland. Meanwhile, in the forests, their somewhat daintier relatives, the mastodons (the size of Indian elephants) fed on a diet of black-spruce boughs and other woody tidbits.

The mammoths and mastodons were relatively recent arrivals on the plains, Ice Age immigrants that migrated across the Bering land bridge from Eurasia during intermittent cold spells. Whenever the climate worsened and the glaciers advanced, water became locked up in the ice and sea levels dropped, exposing a bridge of land across the Bering Strait between Siberia and Alaska. When the glaciers receded again, the land bridge was drowned, but a passageway simultaneously opened to the south through the Canadian plains, which allowed the newcomers to wander into the heart of the continent. Some mammals, including ancestral camels and horses, made this journey in reverse, moving north when the plains corridor was open and then migrating across to Asia when the land bridge appeared.

Columbian mammoth

Of all the species that arrived on the North America plains during the Ice Age—a menagerie that included not only elephants but also grizzlies, elk, and moose—two demand special notice. The steppe bison, *Bison priscus*, was a magnificent, thick-maned animal with flamboyant curving horns (attributes that are dramatically depicted in the cave art at Lascaux, France). The first bison herds likely poured across the Bering land bridge a few hundred thousand years ago and eventually made their way south to the Great Plains. Over the millennia that followed, successive waves of steppe bison made the same long trek, eventually meeting and mingling with the descendants of the pioneer herds. Meanwhile, that pioneer stock had been changing, shaped by life on the steppes and forests of a new continent. The result of this complex process of immigration, adaptation, and interbreeding was the creation of several distinctively North American types, notably the giant, long-horned *Bison latifrons* and the somewhat smaller *Bison antiquus*. In time these species were displaced by an even more compact version, *Bison bison*, the shaggy beast that, in historic times, provided food and shelter to the first people of the plains.

Skulls of Bison
latifrons and
Bison antiquus

Plains bison,
Bison bison

Exactly when the first of those hunting people arrived on the scene is a mystery. Until recently, most archaeologists asserted that humans (members of a genus that was born in East Africa some 2 million years ago) entered North America from Asia, by crossing the Bering land bridge and traveling down an ice-free corridor into the plains. This migration was believed to have happened within the last 11,000 to 12,000 years. Recently, however, archaeologists have uncovered evidence that people were living in Chile at least 1,000 years earlier than this, a discovery that has made the old timelines suspect. But whether the first people arrived 60,000 or 30,000 or 15,000 years ago, over the Bering land bridge or (as some suggest) by boat, as a massed arrival or in staggered groups,

we know that by about 11,000 years ago, they were established across the north-central and southern plains. Equipped with elegantly chipped fragments of stone and bone, these hunters killed and butchered not only bison but also camels, horses, mastodons, and—their specialty—mammoths. At sites from Alberta to Texas, the proof of their presence—blackened hearths, discarded tools, and cracked marrow bones—lies buried where they left it so long ago. In some places, the skeletons of several large mammals lie strewn about the camps, testimony to the success of these recent immigrants.

Three-flowered avens, seen in bud at left, is also known as prairie smoke or old man's beard in tribute to its feathery seedheads, seen here.

But inevitably, on a planet where change is the only constant, their success was fleeting. By 13,000 years ago, the fabulous array of large mammals that the newcomers had encountered on the plains was already disappearing. As many as fifty species—including giant beavers, ground sloths, lions, cheetahs, dire wolves, saber-toothed cats, horses, camels, mammoths, and mastodons—all became extinct within a few thousand years. Did an exploding population of well-fed humans hunt the animals into oblivion, as some archaeologists suggest? Or was climate change the culprit? The evidence suggests that, between about 13,000 and 10,000 years ago, average global temperatures first dropped abruptly and then rebounded. On the North American plains, these climatic changes ultimately translated into a dramatic shift in vegetation patterns. Pushed by warmer, drier conditions, the spruce forests gave way to pines, then in places to open, mixed woodlands, and ultimately to grass. In the blink of an eye (relatively speaking), a carpet of grasses spread out across the plains, blued by sage and beardtongue and enlivened by patches of golden beans, blazing stars, and prairie smoke, or three-flowered avens. The prairies of historic times had finally been created.

This new grassland was big and bold, but it was also much less varied than the mixed landscape of tundra, grass, and forest that it had displaced. And perhaps this in itself is enough to explain the disappearance of the Ice Age megamammals, which required a rich and varied supply of foods that grasslands alone could not provide. Yet if the new habitats were insufficient to sustain mammoths, they appeared perfectly suited to the bison, which soon emerged as the dominant grazing animal on the open range.

As the great herds drummed across the prairie, they of course had no conception of how their homeland had been shaped. The evidence of that drama lay unnoticed beneath their hooves—long-buried evidence of onrushing seas, rising mountains, silt-burdened rivers, and towering cliffs of ice. But the bison were untroubled by the traumas of the past, as they flowed across the horizon toward the present.

three

THE GEOGRAPHY OF GRASS

I am the grass.

Let me work.

CARL SANDBURG, "GRASS," 1918

NATIVE GRASSES ARE the heart and soul of the prairie, the living link between the physical resources of the Great Plains—sunshine, rain, and soil—and almost every other aspect of the ecosystem. At first glance, grasses may look simple or even primitive. In fact, they are highly evolved organisms, especially adapted to cope with extreme climatic uncertainties, including frequent drought. From probable origins in the African region of the supercontinent Pangaea some 60 million years ago, grasses have migrated to every continent except Antarctica and have diversified into about 10,000 species throughout the world. Of these, some 140-odd species, in 41 genera, naturally occur in the Great Plains grasslands. That's twelve dozen distinctly different native grasses! Some of them, like the magnificent big bluestem, or turkey foot (so called for its large, three-lobed seed head), grow up to 10 feet (3 meters) tall. Others, like the stick-in-your-socks specialist needle-and-thread grass, seldom exceed 3 feet (1 meter) in height. At the low end of the scale are species like the diminutive blue grama, which grows close to the ground and rarely raises its elegant, eyebrow-shaped seed heads more than a few hand widths above the soil.

Yet despite these obvious differences, the prairie grasses all share one crucial ability. They are tuned in to the climate, able

to dial their metabolisms down when conditions are unfavorable for growth and speed them up when the weather improves. Far from being passive stalks blowing idly in the wind, prairie grasses are lean, mean growing machines, designed to make the most of limited and unreliable resources.

Managing Moisture

One key to the prairie grasses' success is their ability to conserve water. Like most plants, grasses take in water through their roots and lose it as water vapor through tiny mouth-shaped valves, or stomata, in their leaves. The larger the surface of the leaf and the more stomata it bears, the greater the risk that the plant will lose too much moisture through evaporation and collapse. Grasses are protected from this trauma by having a reduced number of stomata and by the design of their leaves, which take the form of narrow blades. What's more, the surfaces of these reduced leaves are often modified—corrugated with ridges or covered in hairs—so that the wind can't sweep across the surface and draw out moisture. The roughened surface holds a thin layer of humid air next to the leaf and thus helps to reduce the "evaporative demand," or drying power, of the atmosphere. Some grasses, including western wheatgrass, June grass, and blue grama, roll up the edges of their leaves during times of drought to help keep their tissues from drying out.

Why aren't the stomata kept tightly closed to seal moisture inside the leaf? The reason is that the stomata also supply plants with fresh air. Leaves are miracle workers, able to take carbon dioxide from the air and water from the soil, zap them with solar energy, and transform them into food. This process—photosynthesis—not only produces the sugars and other organic molecules that plants need to maintain themselves and to grow but also feeds microbes, worms, insects, fish, birds, and mammals. If plants sealed their stomata, this life-sustaining process would come gasping to a halt for lack of carbon dioxide. But if the stomata are thrown wide open, the plants risk death due to the loss of moisture through their gaping valves.

Prairie grasses resolve this dilemma by strategic scheduling. In the fierce blaze of the midday sun, the stomata close so that water vapor is held in and carbon dioxide is kept out. In this state, the leaf can capture solar energy and store it in energy-rich molecules (a process that requires sunlight but not carbon dioxide). Then, in the cool of the evening, when the evaporative demand drops off, the stomata snap open, letting water vapor trickle out but also permitting carbon dioxide to flood into the leaf. By mobilizing the energy that was

facing page: Although grasses are the most important plants on the prairies in terms of sheer volume, wild flowers (or forbs) also make a vital contribution. With a species list that numbers into the thousands, forbs add diversity to the wall-to-wall carpet of grass, thereby enriching the entire ecosystem. The more plant species are present, the more animals are able to find what they need, whether it be nest sites, daytime shelters, or feeding opportunities. This grasshopper nymph, for example, is taking advantage of the nutritious petals of a prairie sunflower.

stockpiled earlier in the day, the leaf uses this carbon dioxide to manufacture the sugars and other molecules that it needs for growth (a process that can be accomplished in total darkness). The result is that prairie grasses are partially nocturnal; they do most of their growing at night or in the early hours of the morning.

Prairie grasses also have another ingenious way of evading the demands of the sun. Like many other grassland creatures (prairie dogs, ground squirrels, cottontails, badgers, and so on), they take refuge underground. What we think of as "grass"—the aboveground leaves and stems—actually constitutes less than half of the organism. Between 60 and 80 percent of the plant, by weight, typically grows below ground. The roots extend down from the base of the stems like a tangled head of hair, as main roots divide into minor roots and then into root hairs. A 10-foot (3-meter) stand of big bluestem is anchored under ground by a mass of coarse, fibrous roots that reaches as much as a good 12 feet (or 3.6 meters) into the earth. Blue grama, for its part, seldom lifts its seed heads very far above the ground, but its network of fine, branching roots can sometimes probe the soil for water almost 6 feet (2 meters) down!

These extensive systems of roots push thirstily through the soil, intent on sucking up every available drop of water. But if the soil is very dry, as it is during periods of drought, the roots can't draw in enough moisture to keep pace with losses from the stomata. Grasses respond by transferring their most valuable resources (including sugars and proteins) from their leaves into their roots and, especially, into their rhizomes—those aggressive, underground stems that are familiar to anyone who has ever battled with quack grass in the garden. Dead to the world above ground—withered and crisp—the plants live frugally below the surface, drawing on their cached supplies and biding their time until the weather improves. When the rains eventually return, as inevitably they do, the grasses explode into action, sending out fresh rhizomes, which in turn put out fresh leaves and roots, to produce a burgeoning network of tender growth. The amazingly resilient blue grama can revive from dormancy, green up, and grow on as little as .2 inches, or 5 millimeters, of rainfall.

Prairie grasses are not all equally capable of coping with drought. In general, tall grasses, including big bluestem and other shoulder-high species such as switchgrass and Indian grass, require the most moisture, while short grasses like blue grama, galleta, and the stubby little buffalo grass are the most resistant to drought. Midheight species, including needle-and-thread grass, rough fescue,

facing page: Sometimes called turkey foot or beardgrass, big bluestem can be recognized by its distinctive three-pronged heads. The fuzzy hairs on the heads are inconspicuous, scentless flowers.

The ability to survive episodes of drought is a basic requirement for life on the Great Plains.

and western wheatgrass (a.k.a. bluejoint, for its bluish leaf nodes), tend to fall somewhere in between. But all prairie grasses can contend with drought more successfully than can most deciduous trees—which is why the prairies are prairies instead of forests. The grasslands are an expression of the drought-prone prairie climate and a living response to the geography of the midcontinent.

Weather Matters

To the homesteaders who came to the Great Plains from Europe or eastern North America in the late 1800s and early 1900s, converting the prairies to croplands must have looked like a dream. Except for the trees that crept in along the rivers, the land lay open to the plow, offering little apparent resistance to the farmers' ambitions. But the settlers' early optimism was soon blighted by widespread droughts, as the dry summer of 1889 was followed by the dry years of 1890, 1894, 1910, and 1917, and then by the bleak decade of the 1930s. Life on the prairies was not as easy as it had seemed. For what no one at first quite realized was that grasslands are semiarid zones—better watered than deserts but

left: Pincushion cactus

right: Prairie crocus, or pasque flower

> **PLANTS FIGHT BACK**

Prairie plants have come up with many ingenious strategies for coping with water shortage. A few, like the pincushion cactus, are genuinely drought resistant. In other words, they can store water in their own tissues (in their enlarged stems) and draw on it as needed. Others, including many grasses and wildflowers, attempt to evade drought by going dormant and retreating underground, where they linger on in the form of seeds, rhizomes, or tubers. But if some plants favor patient waiting, others put their faith in speed. Instead of trying to sit out the drought, they attempt to avoid it entirely.

Take, for example, the prairie crocus, or pasque flower. An inexhaustible source of pleasure for people on the northern plains, crocuses appear on the trailing edge of winter as tight clusters of furry, pointed buds that push up through the dead grass like so many inquisitive snouts sniffing for spring air. Without pausing to grow leaves, the plants burst directly into bloom, producing ground-hugging whorls of silky, lavender sepals. By the time most other wildflowers put in an appearance several weeks afterward, crocuses are already sporting headdresses of shiny, plumed seeds. Before the growing season has even properly begun, their reproductive task has been completed.

By getting off the mark so early, crocuses are able to draw on a relatively certain supply of water from snowmelt. And although they are exposed to the bluster of winter's last blast, they are protected from the wind by a coat of hairs that holds in heat and moisture. They also take shelter by crouching close to the ground, well bedded in grass, creeping juniper, and other plants. Thus protected, crocuses speed through their reproductive cycle and avoid the stress of coping with drought in the hot, dry days of July and August.

less humid than forests. The farmlands that the settlers had known in Europe and the East had typically been wrested from the forest and, even after the trees were gone, still received enough rainfall to support a natural vegetative cover of broad-leaved woodlands. But the weather on the prairies naturally favored not trees but grass, and that simple fact made all the difference.

Like most of the world's great grasslands, the Great Plains of North America lie squarely in the middle of a large continental land mass. As a result, the region is isolated from the influence of all four oceans—north, south, east, and west—and, as it happens, from any other significant body of water. Without the moderating influence of water (slow to heat and slow to cool), the plains are subject to violent oscillations of temperature. In the northern prairies, in particular, the temperature can span 140°F in the course of a year, from a brittle −40°F in midwinter and to a stifling 100°F in summer. (That's a range of 80°C, from a low of around −40°C to a high of over +40°C.) The effect of these wild seasonal deviations is equivalent to moving up and down the continent every twelve months. Saskatoon, for example, has an average January temperature of 0°F (−17°C), well below that of Anchorage, Alaska. But in July, Saskatoon's average heats up to 66°F (19°C), almost on a par with that of Los Angeles. And though the southern plains are spared the worst extremes of winter, they still get taken for quite a ride. The average January temperature in Amarillo, Texas, for example, is a mere 35°F, or 2°C, cooler than Vancouver far to the north; but in July, the north Texas

Indian grass

Galleta

Needle-and-thread grass

Western wheatgrass

Buffalo grass

Sideoats grama

Little bluestem

Several million years ago, perhaps about the time when miniature rhinos and dog-sized horses were roaming across the prehistoric savannas of the Great Plains, life was confronted by a major challenge. For some unknown reason, the concentration of carbon dioxide in the atmosphere had begun to drop—so much so that certain plants were left gasping for breath. Without adequate carbon dioxide, they could not easily carry out the life-giving process of photosynthesis.

Some of the afflicted plants, including several families of grasses, responded to this crisis with a major innovation, the invention of a completely new method of photosynthesis. Unlike the old, so-called C_3 pathway, which had chugged along for millennia by linking carbon atoms in three-atom molecules, the new, improved C_4 model produced more carbohydrates from the same amount of carbon dioxide by linking carbon in four-carbon molecules. But to achieve this enhancement, the plants relied on an enzyme that required relatively high temperatures, with an ideal range between 75°F and 85°F (25°C and 30°C), up from the old standard of 65°F to 75°F (20°C to 25°C).

Although carbon dioxide levels eventually recovered from their slump, C_4 photosynthesis proved to be so successful that it has survived to the present day, alongside the tried-and-true C_3 technology. Today, C_4 grasses—including blue grama, sand dropseed, big and little bluestem, and corn—are found throughout the prairies and are especially dominant on the southern plains (where heat and moisture are abundant at the height of the growing season). Because of their temperature requirements, they generally grow and set seed in mid to late summer and hence are known as "warm-season" grasses. The C_3, or "cool-season" grasses, by contrast—including needle-and-thread, western wheatgrass, and wheat itself—typically produce seed early in the summer before the temperatures hit their peak. Since in nature C_3 and C_4 grasses typically occur in mixed stands, they are able to share scarce water by pulsing their demands.

plains can be among the hottest places on the continent. (If prairie people are obsessed by the weather, it may simply be because we have a lot of weather to obsess about!)

Prairie grasses ride this climatic roller coaster with composure. Species that couldn't stay the course fell into extinction in ages long past, leaving behind the modern community of hardy survivors. Each of these successful species has been further refined over the past several thousand years, producing sub-species, or varieties, that are finely attuned to local conditions. A native grass from Alberta, for example, typically achieves maturity in a matter of weeks, fitting its life cycle to the abbreviated growing season of the northern plains. But a clump of the same species from Missouri or Oklahoma is programmed to take its time, pacing its activities to the more leisurely schedule of southern climes. These kinds of local, genetic differences have been detected in a wide range of native grasses, including blue grama, its cousin sideoats grama, the

facing page: Winter on the prairies is a force to be reckoned with.

compact-and-graceful June grass, switchgrass, and both big and little bluestem. Natural selection, that master gardener, has been at work on them.

The one climatic factor that presents a continuing challenge for prairie grasses is the moisture supply. Over the Great Plains as a whole, precipitation is more variable than it is almost anywhere else on the continent, with years that are both much wetter and much drier than the long-term norm. (One study of precipitation records in western Kansas, for example, showed that in most months the amount of moisture received was either significantly below or significantly above average. Only "normal" values were truly abnormal.) When the rains are generous, the prairie flourishes and blooms; but when drought sets in, the grasses—indeed the whole ecosystem—are severely tested. More than any other single factor, the limits to growth on the prairies are set by precipitation.

Why So Dry?

The Great Plains are subject to drought partly because they lie in the lee of the western mountains. Without this elevated barrier, westerly winds from the Pacific could sweep across the plains and bring moisture to the dry lands from Airdrie to Abilene. But with the Coast Ranges, the Cascades, the Sierras, and the Rockies all standing in their path, the Pacific westerlies are forced to rise, cool, and drop their moisture as they pass. By the time the winds flow down over the plains, they are almost devoid of rain. As they move across the western grasslands, they pick up humidity and carry it to the well-watered eastern forests. (Sometimes, as the winds swoop over the mountains, they whip themselves up into disturbances—known as Alberta Lows or Colorado Cyclones, depending on where they occur—that sweep across the plains, carrying a splattering of rain or a dusting of snow.)

Once the westerlies get past the mountains, there is nothing to stop them— except for the invisible resistance of other air masses. The prairies' wide open spaces are a playground for the winds, drawing in not only mild, dry air from the Pacific but also colder and even drier air from the tundra and polar seas. As this Arctic air floods south, it meets moist, tropical air flowing north from the western Gulf of Mexico. A typical weather diagram for the Great Plains would show Arctic air pushing down from the north, tropical air swinging in from the south, and wrung-out Pacific air wedged between them like the point of an eastward-facing arrow. Arctic air dominates in the winter, sometimes forcing itself all the way to Texas as a stinging "blue norther" and, occasionally, pushing

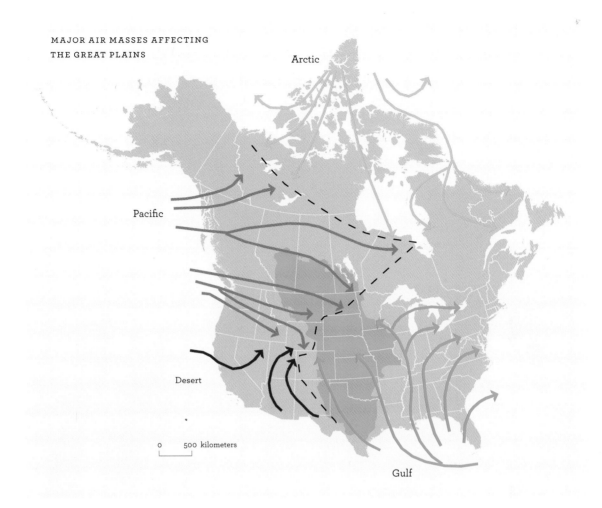

MAJOR AIR MASSES AFFECTING
THE GREAT PLAINS

Arctic

Pacific

Desert

0   500 kilometers

Gulf

on south to the isthmus of Tehuantepec at the very tip of Mexico. In the spring, the balance of power is reversed, as the gulf air mass gains strength and surges north, sometimes carrying tropical heat and humidity all the way up to the Canadian prairies. The rained-out Pacific westerlies, with their meager stock of moisture, make themselves felt throughout the year, especially during the dry months of fall and winter.

In spring and summer, in particular, this picture is complicated by an influx of warm, dry-as-bone air that blows across the southern plains from the southwestern deserts. The interplay of these "four strong winds" produces the distinctive precipitation patterns of the Great Plains grasslands. For instance,

For a person who craves the excitement of "active weather," there is no better place to be than out in the middle of the broad prairie. The violent interplay of winds from the west, north, and south, complicated by occasional blasts of desert air from the southwest, all sheering across each other at their own heights and speeds and brought to a boil by the summer sun, builds up into magnificent thunderheads that rumble and spit bolts of fire across the grasslands. Not infrequently, these storm centers grow into supercells that pelt the Earth with hail, a hazard that can strike anywhere but that is especially prevalent in west-central Alberta and the central U.S. plains, in a large arc round and about Cheyenne, Wyoming.

When thunderstorms get especially riled up, they occasionally produce whipping tentacles of wind that, more occasionally yet, reach down to the ground and churn across the landscape as tornadoes. Although always rare and exceptional, tornadoes are nonetheless much more common in North America than anywhere else on Earth, with most reports coming from the southeastern United States and the Great Plains. Among the regions with the highest incidence of twisters is "Tornado Alley" in the southern plains, a zone that extends from Texas and Oklahoma, into western Missouri, and north through Aunt Em's farm in Kansas.

winter is a relatively arid season across most of the plains not only because cold air cannot hold much moisture but also because of the strong seasonal influence of dry air from the north and west. As a rule, less than one-third of the year's precipitation falls between October and March, when these air masses exert their strongest influence. The other roughly 70 percent of the year's moisture is received during the April-to-September growing season. Without this well-timed gift, the Great Plains would be a prickly expanse of cactus and other desert plants.

Much of the all-important spring-and-summer rainfall is generated when tropical air surges north and runs into Pacific and Arctic systems moving across the prairies to the east and south. Where the air masses collide, the lighter, warmer air from the tropics is forced up, cooling as it climbs, condensing to form clouds, and ultimately losing its moisture as general rain showers. Violent thunderstorms also frequently develop along these collision zones, or fronts, as the unstable tropical air rises up into towering, super-energized cumulonimbus clouds that glower over the landscape before releasing their humidity as hail or pounding downpours.

Because the contending air masses often meet in midcontinent, frontal thunderstorms are most common in the middle of the plains, in and around Colorado, Wyoming, South Dakota, Nebraska, and Kansas. But storms can also develop locally, without the clash of opposing weather systems to set them off. All it takes is a mass of warm, moist air and something to send that air spiraling up through the atmosphere. This lift-energy usually comes from the summer sun, which blazes down through cloudless prairie skies to heat the ground. Heat then radiates out of the soil into the surface air, causing it to rise, rotate, and mount upward to form a rain-filled thunderhead. In the dry western plains, one-third of the year's precipitation can fall in a single hour from one of these spectacular cloudbursts.

facing page: A severe thunderstorm crackles over the hills of Grasslands National Park, near Val Marie, Saskatchewan.

Global "Teleconnections"

By and large, the prairie climate is reliably unreliable. As the rival air masses interact with each other over the plains, they keep the atmosphere in a state of more-or-less-constant flux, so that the weather oscillates from extreme to extreme. But there are also times when the climatic system seems to get stuck. "Wet spells," for example, when the rain refuses to stop. "Dry spells" of months— or years—when the clouds seem dry as parchment and the air fills with dust.

These persistent weather patterns also tend to be widespread, affecting significant parts of the Great Plains for prolonged periods. The droughts of the 1930s, for example, occasionally flared out to singe the entire continent, but they were at their most intense across the Great Plains grasslands. Some parts of the High Plains in Oklahoma and Texas experienced eight consecutive years of drought, between 1933 and 1940. Then, little more than a decade later, the central and southern plains—from the Mississippi to the Rockies and from Colorado to Texas—were again stricken by a severe drought that persisted from 1952 to 1957. The Canadian Prairie provinces were hit hard in 1961. Then, in the late 1980s, a three-year drought parched the entire northern plains and fuelled disastrous forest fires in Yellowstone National Park. During the growing season of 1988, when the crisis was at its worst, many parts of the prairies were hotter and drier than they had been at any time during the Dirty Thirties.

Yet five years later, some of these same areas were in full flood, as torrential rains pounded the western Midwest and sent both the Missouri and the upper Mississippi rivers spilling over their banks. By the time the waters receded, twenty-six people were dead.

Why do the prairies suffer these violent climatic spasms? Part of the answer to this question may lie halfway around the world, in a region somewhere between Australia and Peru. There, in the equatorial waters of the South Pacific Ocean, weather patterns that will eventually affect the prairies begin to brew. Recent research suggests that there is a link between the surface temperature of the South Pacific and the amount of precipitation that falls on the Great Plains, particularly during the winter and early spring.

So far, no one knows exactly how all the complex linkages in this global "teleconnection" work. But by rushing in where angels fear to tread, we can sketch a plausible, if simplified and speculative, narrative. When surface currents in the equatorial South Pacific are unusually warm—during a so-called El Niño event—huge volumes of water evaporate from the ocean into the air. When this massive upwelling of warm, moist air spirals into the atmosphere, it stirs up the circulation so energetically that the airflow over North America is affected. As the air masses swirl into new alignment across the continent, a persistent low-pressure system may eventually develop somewhere over the southern Great Plains. Since air flows toward areas of low pressure, this Low acts as a kind of pump that sucks in humid air from the gulf. Moist and warm, heavy with rain, this tropical influx delivers tons of precipitation onto the southern plains.

Farther north, on the Canadian prairies, the effects are reversed, as El Niño heralds unusually mild winters and dry spring weather. But when the surface currents in the equatorial South Pacific are exceptionally cold—when La Niña sets in—the northern plains are likely to be frigid and buried deep in drifts.

But global influences, however stupendous, are not the only factors at work. Often, extreme conditions linger on the Great Plains long after the systems that triggered them have dispersed. A wet spell seems to breed more wet weather; a dry spell appears to breed more drought. But how could weather patterns possibly perpetuate themselves? The answer turns out to be surprisingly obvious. When precipitation is plentiful, water accumulates in the soil. As plants draw on this moisture to grow, they release water vapor into the air. This water vapor, in turn, combines with humidity that has evaporated directly from the earth, and these exhalations rise together to form clouds. Thus rain in the soil begets rain showers. What's more, both rainfall and evapotranspiration (the release of water vapor from plants) have a cooling influence that helps to moderate temperatures and keep the evaporative demand within comfortable limits.

But following a prolonged dry spell, the cycle grinds to a stop. Plant growth slows and the rate of transpiration declines. So too does cooling evaporation

The cruel-looking spike on the rear of this female Mormon cricket is actually an egg-laying organ, or ovipositor. Although she does her best to protect her eggs by burying them underground, the outcome really depends on the weather. A cool, wet season means a poor hatch, while a warm, dry spring brings the reward of numerous offspring.

Candace Savage photo

from the soil. The ground and the surface layers of air sizzle in the sun, as a hot, dry land gets hotter and drier. (A case in point is the drought of the 1930s, which seems to have been intensified and prolonged by farming methods that left the soil exposed to the parching wind and robbed the system of what little moisture it held.) Eventually humid air from the south or the west returns to the scene, bringing welcome relief and restoring the climate to its own eccentric sense of normalcy.

Gradients of Grass

These prolonged episodes of drought (many of them "forced" by events halfway around the world) have been the making of the Great Plains grasslands. Drought sucks moisture out of the soil, beginning at the surface and gradually burning farther down. If a dry spell is brief, the deep stores of moisture remain untapped, but if the evaporative demand persists, even the subsoil becomes parched and cracked. As a result, deeply rooted trees can cope without rain for several years by drawing water from underground, but they are doomed to

> ### YOU CALL *That* A DROUGHT?

The Great Plains grasslands—and the climate that defines them—have been around for the last 8,000 to 10,000 years, ever since the last retreat of the glaciers. In the early days of this regime (from about 8,000 to 4,000 years ago), the climate was considerably warmer and drier than it is today and even more prone to drought. But sometime in the last few thousand years, the system took a turn toward cooler, moister norms, so droughts have ever so gradually and erratically become less frequent.

In fact, it seems that the past century (from 1900 to 2000) was the wettest in 2,000 years. This conclusion is based on studies of microscopic fossils found in lake beds across the northern plains, in Alberta, Saskatchewan, and North Dakota. By extracting core samples from lake bottoms and studying the types of fossils that are found at different depths, researchers are able to estimate the salt content of the water at different times in the past. Since salinity increases as water levels drop, these findings give them a measure of past droughts. At Humboldt Lake in central Saskatchewan, for example, the fossils bear mute witness to a severe drought—as bad as that of

1988—that persisted unbroken for more than seven decades. (On the southern plains and in the desert U.S., researchers have uncovered evidence of prehistoric droughts that persisted for up to 300 years.) Overall, it is clear that the droughts of the last century have been nothing compared with the intense and prolonged aridity of preceding millennia.

Could the prairie climate someday revert to its fierce old habits? Climatologists say that this is not only possible but quite likely. Scenarios for global warming (the anticipated shift in the climate due to the greenhouse effect) suggest that the Great Plains will experience higher summer temperatures and more frequent and extreme droughts, and that these effects will become increasingly evident within the next fifty years. Already, ecosystems around the world are showing signs of change, with earlier dates for flowering, emergence of insects, and egg-laying. And across the northern and central Great Plains, average annual temperatures are on the rise—up about 2.2°F (1.2°C) in some localities, in the last fifty to a hundred years—as the prairies ride the leading edge of global warming.

defeat when drought reaches their root zone. Meanwhile, the grass lies patiently around the dying trunks, ready and able to spring back to life when the new rains finally come.

Long-term patterns of precipitation not only determine whether the land will grow trees or grass but also establish the limits that distinguish one "type," or ecoregion, of grasslands from the next. When precipitation and other variables are averaged over the long term, the underlying order of the prairie climate begins to emerge. In the textbooks, these hidden patterns are revealed through charts and maps, but out on the prairie, they are written as gradients of grass.

Sometimes, the dialogue between the vegetation and the climate is intriguingly complex. For instance, summer precipitation on the prairies depends, in large part, on air masses that blow in from the south, carrying moisture from the gulf. Because of their southerly origins, these winds naturally have a greater influence on the southern plains (where they "reside") than in the north (where they merely "visit"). So it isn't entirely surprising to discover that the southern plains receive significantly more moisture than the northern prairies do. If, for example, Amarillo can hope to get 20 inches, or 500 millimeters, of moisture in a normal year, Lethbridge typically has to make do with only three-quarters as much. With this difference in mind, one might expect the prairies of northern Texas to be lusher than those of southern Alberta or Saskatchewan. Instead, the reverse is the case.

The trick is that the south-to-north gradient in precipitation is canceled out by an equal but opposite north-to-south gradient in evaporation. Because the average annual temperature increases from north to south, so does the rate at which moisture is lost through evaporation. Whatever the southern plains gain as rain, they lose as water vapor. As a result, the "effective precipitation"—the amount of water that is available to growing plants—is about the same in

Taking full advantage of a well-timed spring rain, Texas bluebonnets and Indian paintbrush blaze along a roadside near Llano, Texas.

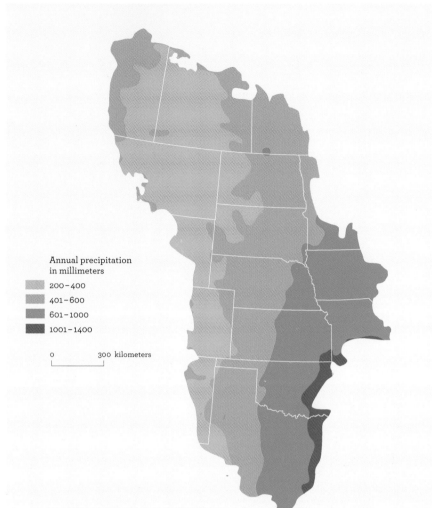

Annual precipitation
in millimeters

200–400
401–600
601–1000
1001–1400

0   300 kilometers

AVERAGE ANNUAL PRECIPITATION
ON THE GREAT PLAINS

facing page:
Controlled burning
is used to maintain
the health of tall
grasslands, like this
one in Badlands
National Park,
South Dakota.

southern Alberta as in northern Texas. This helps to explain the long, gradual transition from the semiarid climate of the Northwestern Short/Mixed Grasslands to the sun-frazzled conditions of the Southern Short Grasslands. (A map showing these ecoregions can be found on page 22.)

Meanwhile, there is yet another climatic gradient that also helps to shape the vegetational profile of the Great Plains. This is an east-to-west decline in average annual precipitation. The tropical air that brings summer rains to the prairies typically swings up from the Gulf of Mexico, through the central

> PRAIRIE FIRE

Climate is the major factor that determines the extent of the Great Plains grasslands. To put it in technical terms, grasses hold sway wherever the evaporative demand (the amount of moisture that the atmosphere would draw away if it could) is slightly greater than the precipitation (the amount of moisture that is actually out there, in the ecosystem). But there is one important exception to this rule of thumb. The lush tall-grass prairies that fringe the eastern margin of the plains receive abundant moisture, more than enough to keep pace with evaporation. Theoretically, then, the region ought to support trees. And, in fact, wherever fragments of native tall-grass prairie have survived untilled, they have been aggressively invaded by stands of aspen, oak, and dogwood during the last 150 years.

The missing link—the force that maintained the tall-grass prairie in the past but that has since been all-but-extinguished—is fire. Prairie fire was the terror of the early settlers, a rapacious enemy that raced across the plains, leapt over firebreaks and rivers, and at times sped from horizon to horizon in a single day. But alarming as it was to humans, wildfire was a friend and ally of the tall grasses. Not only did it clear away the thick thatch of dead vegetation that prevented new shoots from breaking through, it also killed trees, the true "terror" of the prairies. When a tree burns, the growth points on its twigs and branches are likely to be injured, so the plant is hampered in its ability to produce new shoots. But a grass protects its growing tips under the ground, where the fire cannot reach, and rises from the flames like the proverbial phoenix.

Before the agricultural era, most parts of the tall-grass prairie were probably set ablaze once every three to ten years, either by lightning or by the deliberate action of Native people, who used fire to green up the prairie and bring in animals. But however prairie fires started, they always had the same effect: they were the guardians of the tall-grass prairies.

Raindrops cling to the translucent petals of a harebell, a flower that is native not only to the northern Great Plains but to parts of Europe as well, where it is famous as the bluebell of Scotland.

United States, and off toward the east. As a result, its influence is stronger on the tall grasslands of the eastern plains than on the short-to-mixed grasslands farther west. If Winnipeg receives about 20 inches (500 millimeters) of moisture on average, Lethbridge gets 20 percent less (just 16 inches, or 400 millimeters). And if Kansas City can count on almost 40 inches (1 meter) of precipitation in a normal year, Amarillo can only expect to receive about half as much—and this time there is no reverse gradient in temperature to compensate for the difference. Less moisture is simply less.

In the days before the prairies were plowed and settled, this east-west moisture gradient found expression in the natural vegetation. As the average precipitation declined toward the west, the vegetation diminished in step, gradually reducing in height like a living bar graph. Somewhere around 100 degrees west longitude (give or take a few degrees), the tall-grass species dwindled away, leaving the wind to ripple through knee-high stands of mixed-grass prairie. To the west and southwest, the midheight grasses in turn gave way to a carpet of ground-hugging grasses, as a dry land clothed itself in drought-resistant plants.

The boundaries between tall-, mixed-, and short-grass prairies are not as tidy as they look on the map. In the patchwork quilt of the grasslands, each of the major blocks of vegetation is composed of many smaller blocks. And just as the overall picture is determined by large-scale climatic patterns, so each of these distinctive patches is a response to local variations in microclimate. Conditions are different on a south-facing slope than on the north and on lowlands than on hilltops, and these subtle differences are reflected in the vegetation. Moving uphill, from humid bottomlands to the drier crest, reproduces the moisture gradient of moving from east to west. And so, a sea of tall-grass prairie is broken by islands of mixed (or even short) grasses that grow on uplands and arid slopes. Meanwhile, out on the short-grass prairie, blue grama and its diminutive associates follow the opposite trend, ceding ground to midheight or tall grasses in moist valley bottoms.

Whether considered locally or regionally, the transition zones between the different "types" of prairie are dynamic, migrating upslope and down-slope, east and west, following changes in the climate. During periods when the weather is unusually wet, the tall and mixed grasses expand their influence and range, tracking the sweep of abundant rain. In times when the prairies are stricken by drought, the tall grasses are forced into retreat and the drought-resistant short grasses extend their reach. (During the Dirty Thirties, for example, the short-grass community pushed as much as 150 miles, or 240 kilometers, to the east, temporarily usurping the tall-grass prairie.) Thus, the grasslands keep time with the climate, mirroring its ebbs and flows and dancing to the erratic beat of the rainfall.

Western wood lilies scarcely put in an appearance during dry years but burst into a riot of color when the rains return.
Arthur Savage photo

four

SECRETS OF THE SOIL

…the prairies are nothing but grass
as the sea is nothing but water.

WILLIAM LEAST HEAT-MOON, *PRAIRYERTH*

THE GRASSLANDS OF the Great Plains tug at our senses, challenging us to overcome our limitations as mere humans. On the one hand, the landscape calls us to enlarge our horizons and soar across its spacious expanse. But on the other, the prairie seduces us with details, quietly drawing our attention to the little things that matter in this big, round world. The flick of a ground squirrel diving into its burrow, the wheeling shadow of a hawk, the ripple and run of its wingspan across the bending grass. If we are willing, the prairies will even lead us beyond these small sightings to truly minuscule events, beckoning us into an alien but all-important world that lies almost entirely beyond the reach of our ordinary awareness. For the typical life-form on the grasslands is not a bison or a gopher or even a blade of grass. It is some weird little creepy-crawly, far too small to be seen by the unaided eye, that lives and dies just beneath our feet in the deep prairie soil.

Recent estimates suggest that the total weight, or biomass, of the invisible organisms that live in prairie soils is greater than the mass of all the visible, aboveground animals put together. If all the microflora and microfauna from grassland soils could be piled onto one arm of some Great Cosmic Scale, with all the insects, reptiles, mammals, and birds from the upper world heaped onto the other pan, the balance would tip sharply toward the soil organisms. Together with the plant roots around which they live, these little creatures are the powerhouse of the prairie, responsible for anywhere between 60 and 90 percent of all the biological activity in the Great Plains grasslands.

Entering the Underworld

The soil has been described as the "poor person's tropical rainforest," one of the richest and most diverse ecosystems anywhere on Earth. The most numerous organisms in this busy world are also the smallest—thousands upon thousands of different kinds of single-celled organisms, including algae, fungi, bacteria, and protozoans. Although a few of these creatures may be familiar from Biology 101—the ooze-along, blob-shaped amoebae, for example, and the whip-tailed flagellates—many (if not most) of the soil microbes are still completely unknown, never having been studied, named, or even seen by anyone. A single teaspoon of dirt typically holds around 5 billion of these fantastically varied little animals and plants, all of them fiercely engaged in the business of life and death. In their billions and their trillions, they expend as much energy as if several thousand people were living on every acre, or hectare, of the prairies.

Waterbear

Adding to the hubbub is a lively assemblage of somewhat larger (though still mostly microscopic) and more complex animals. The bestiary includes plump, lumbering, eight-legged little critters called tardigrades, or waterbears, that waddle through the film of water around soil granules, searching for food. When they find plant roots or fungi, the "bears" drill into the cells with their sharp snouts, or stylets, and suck out the nourishing fluids. Another, more important group of plant eaters, the springtails spend their lives bumbling blindly through the dirt, munching on bacteria, fungi, dying roots, and the occasional dead waterbear or worm. Jumpy and alert, springtails are equipped to leap into action at the first tremor of danger. As their name suggests, they have spring-loaded tails that when all is well they carry tucked up under their abdomens. But when danger is detected, the tail fills with fluid and suddenly snaps down, sending the springtail catapulting to what, with luck, is safer ground. (The light-colored specks of "dust" that sometimes somersault away from compost when the pile is turned are actually springtails taking flight from the disturbance.)

About one square yard of prairie can support anywhere from 5,000 to 15,000 springtails in the top foot of soil. (That's about a square meter dug to the depth of 30 centimeters, or enough to make a couple of wheelbarrow loads.)

But these statistics pale compared with the totals for another major group of soil dwellers. The soil is crawling with microscopic mites, those scurrying, scuttling, eight-legged cousins of spiders and ticks. The same small patch of prairie soil that is hopping with springtails may simultaneously be home to up to 60,000 mites, representing dozens, or even hundreds, of species. (Although 48,000 kinds of mites have so far been identified around the world, the final catalog of species is expected to reach half a million or more.) While many mites make their living peaceably by feeding on microbes and plant roots, others are voracious predators—the tigers of this invisible world. *Spinibdella cronini*, for example, is a largish red mite (about the size of the period at the end of this sentence) that is common in prairie soils and that preys on soft-bodied mites and springtails. Its modus operandi is to seize its prey by surprise, tie it down with silk, and then suck the life out of it.

The opulent, fragrant blossoms of the gumbo, or tufted, evening primrose are typically found on dry, eroded slopes, usually where shales laid down in the Cretaceous Period are near the surface. These insect-friendly flowers open in early morning and close at night.

Pity the poor springtails. For "giant" mites are not the only menace in their shadowy world. They must also be prepared to flee from pseudoscorpions, jaunty little monsters that stalk through the soil with their venomous pincers armed and ready for action. Danger can also come swimming through the darkness in the form of carnivorous worms, innocuous-looking creatures (mostly smaller than the letter i) that either swallow their prey whole or attach themselves to the victim's body and scrape at it until the internal organs can be drawn out through the puncture. These silent predators are members of an enormous tribe of roundworms, or nematodes, that are the most numerous multicellular animals on the grasslands. Hundreds of species have been found at sites across the Great Plains, including those that feed on bacteria, protozoans, fungi, algae, mites, other nematodes, and the roots of higher plants, or some combination of the above. A heap of soil that contains thousands of springtails and mites typically also holds between 3 million and 5 million nematodes.

Nematodes are not only the most numerous multicellular animals on the grasslands. They are also the most important herbivores. While aboveground plant eaters nibble on stems and leaves, herbivorous nematodes eat their way through the roots, attacking silently from beneath. One study conducted on a conservation reserve determined that the unassuming little worms worked their way through more plant matter per unit area than either bison or prairie dogs. And on pasturelands, the standing population of nematodes has been known to consume twice as much vegetation as a herd of cattle.

Fortunately, some plants are able to protect themselves against the attack of the killer worms. A number of common prairie forbs, including pasture sage (or fringed sagewort) and nodding onion, produce natural nematicides that ooze out of the roots and create a worm-free zone for themselves and their neighbors. Such protection is presumably one of the benefits of a diverse plant community in which the roots of different species are intertwined.

But not all plants are so community minded. Some use chemical warfare not to fight against pests but to resist the encroachment of other plants. For example, a dainty-looking short-grass species called prairie three-awn produces root exudates that kill nitrogen-fixing algae and bacteria. (These are soil organisms that capture nitrogen gas from the air and convert it into a water-soluble form in the soil, ready for plants to use.) Because prairie three-awn can get by without much nitrogen, this action creates ideal conditions for its own growth and keeps out competing species that need nitrogen-rich soil.

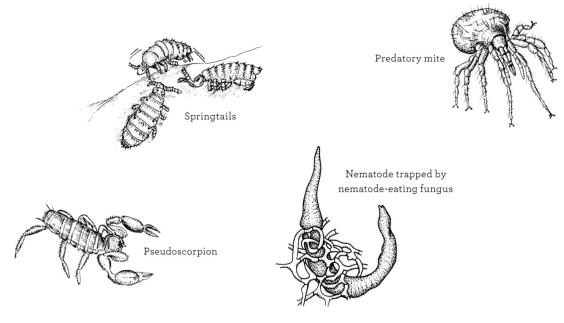

Springtails

Predatory mite

Pseudoscorpion

Nematode trapped by nematode-eating fungus

Across the broad expanse of the grasslands, their numbers spin out of control, quickly mounting toward the astronomical.

Growing Soil

All soil is full of life, whether under prairie, forest, tundra, or desert. Yet compared with other biomes, the temperate grasslands are remarkable for the extent to which they concentrate their resources in the underworld. The reason is simple: it's safer down there. The prairie climate is grueling— too hot, too cold, too wet, too dry, too wild and variable—but go a little way beneath the surface and the environment becomes surprisingly equable. During the growing season, for instance, when the thermometer can easily spike from 50°F to 105°F (10°C to 40°C), the temperature 20 inches (half a meter) down in the soil remains pleasantly cool and virtually constant.

In a region where moisture is often the limiting factor for life, the soil also provides a buffer against drought. Long after the surface of the ground has been wicked dry by the sun and wind, the soil retains hidden reserves of moisture. Not only does water pool in the spaces, or pores, between the soil granules, but it is also glued to the surface of the particles by powerful chemical bonds. Thus, even apparently dry dirt often retains a thin film of water that is tightly bound to the soil grains. To tap this precious resource, grasses put down dense networks of thin, fibrous, probing, branching, fast-growing roots, all seeking moisture and nutrients. On average, grasses produce three or four times more roots by weight than they do leaves and stems, giving them a root-to-shoot ratio that is ten times as high as that of a forest.

The sheer volume of roots produced by prairie grasses is mind-boggling. In the mid-1930s, for example, one determined agrologist undertook to excavate and measure all the roots produced by a small patch of tall-grass prairie in Nebraska. From half a square yard, or meter, of ground, he extracted about 150 miles (250 kilometers) of fibrous roots—enough, laid end to end, to reach

The claret cup, or mound hedgehog-cactus, is a dryland specialist that prefers well-drained gravelly or sandy soil. A native of the southwestern deserts, it can also be found in suitable sites on the southern plains.

Many of the microorganisms in the soil can infect plants and sap their strength, even to the point of making them visibly "sick." But there are also remarkable cases in which infection produces health, as plants and microbes join forces for their mutual benefit.

For instance, almost all prairie plants depend on mycorrhizae, or fungal cells that infect root tissues and feed on the plant's hard-won reserve of sugars. As much as 30 percent of the plant host's resources may be spent on feeding its fungal guest. In return, the fungus produces a network of ultrafine threads that extend out from the roots and penetrate into the tiniest interstices of the soil. A pinch of soil that contains an inch or so of roots and root hairs might also hold up to 130 feet (40 meters) of fungal filaments. This extensive network feeds water and nutrients back to the plant, amply reimbursing it for its services.

An even more remarkable collaboration occurs between the roots of certain plants—including alders and many legumes (vetches, milkvetch, vetchlings, clovers, and the like)—and a select group of nitrogen-fixing bacteria. In this case, the plant often takes the initiative by producing flavorful chemicals that draw the appropriate species of microbes to the site. Once contact has been made, the bacteria induce the plant to open a tubular passageway into the root so that the invaders can move unhindered into the underlying tissues. Not only does the plant feed the bacteria (again at great expense), it also shelters them in sturdy, pea-sized chambers, or nodules, on its roots.

The payoff for this effort is immense. The bacteria in the nodules extract nitrogen gas from air pockets in the soil, metabolize it, and convert it into ammonia. In this form, the nitrogen becomes available to plants, which use it to manufacture proteins. Although a certain amount of nitrogen is fixed by organisms that live freely in the soil, the bulk of nitrogen fixation is performed by bacteria housed in nodules. Without this delicate symbiosis between microbes and plants, life as we know it could not exist.

Purple prairie clover

Golden bean with root nodules

from Lincoln, Nebraska, to Sioux City, Iowa. The tally included 11 miles (17 kilometers) of green needlegrass roots, 13 miles (21 kilometers) of big bluestem, 23 miles (37 kilometers) of little bluestem, and 109 miles (176 kilometers) of Kentucky bluegrass. And that was without factoring in the silvery fuzz of absorbent root hairs that coat root fibers from base to tip. In another heroic experiment, a single rye plant (the domesticated cousin of several native prairie grasses) was found to have produced 7,000 miles (11,000 kilometers) of roots and root hairs in one four-month growing season.

In nature, these root systems obviously do not strike out across the continent. Instead, they squiggle and writhe into every nook and cranny of the soil, holding the dirt together with their slender, grasping fibers. Since most prairie grasses also produce lateral rootstocks, or rhizomes, that send down roots along their length, each plant—and each clump of root-bound earth—is connected to the next and the next. As plant intertwines with plant, and roots interweave with roots, the soil becomes tightly tied together in a thick, fibrous mat—the famous prairie sod, which the settlers used for building their first homes.

For thousands of years before the settlers arrived, that same deep fabric of roots had served as the foundation of the soil ecosystem. The enormous biomass of underground plant matter, both living and dead, provides an almost inexhaustible supply of food for soil organisms. Huge populations of herbivores graze on the roots themselves, including hordes of fungi, nematodes, and springtails. Still other microorganisms feed on the rich soup of nutrients that leak from the roots into the dirt. (Since these resources are concentrated in the rhizosphere—the soil that surrounds the root—much of the life in the soil is concentrated there as well.) In the dark underworld of the soil, everything gets eaten. The dead, the dying—even minute particles of excrement—are all on the menu for some miniature soil creature.

As a result of this endless round of digestion, organic material in the soil (principally dead roots) is first broken down and then gradually built back up into complex and relatively stable molecules known as humus. Dark brown or black, these humic substances not only give prairie soil its rich palette of tones but also contribute to its ability to support a dense tapestry of crops and native grasslands. To thrive, plants need to take in nutrients such as potassium and nitrogen (the K and N found in synthetic plant foods), which they soak up through their roots. Because these chemicals are water-soluble, they are readily leached down into the earth, beyond the depth at which the plants can reach after them. But if the nutrients are bound up in humus, they are effectively stored in the upper 6 inches to 1 foot (15 to 30 centimeters) of the soil. As the soil organisms eat away at the hard-to-digest humus, the stored nutrients are gradually supplied to the plants, much as if they were being freed from a time-release capsule.

Thus, prairie soils are fertile largely because—from the end of the Ice Age until the land was plowed—the native grasses have consistently produced more root fibers than the soil organisms have been able to consume. The surplus resources have collected in thick layers of dark, crumbly, humus-rich earth. (By contrast, desert soils contain little or no humus and forest soils have only a thin, upper horizon, or layer, of black dirt.) To an earlier generation of soil scientists, these remarkable grassland soils were known as "prairyerths," simply and profoundly because that is what they were.

Formative Influences

Technically speaking, prairie soils are characterized by a deep surface layer, or A horizon, of topsoil; a lighter-colored B horizon of subsoil; and a band of calcium-rich salts somewhere within about 6 feet (2 meters) of the surface. But within the broad bounds of this definition, there is room for almost infinite variability. Soil is a unique and intimate blending of living and nonliving elements, and although grasses have doubtless been the major influence on the development of prairie soil, they have not acted alone. The physical setting of the Great Plains has exerted a powerful force of its own.

Just as the sea can be described as a body of water, so the soil could be defined as a body of weathered rock that, like the sea, is permeated by life. The nature of the soil ecosystem depends, at least to some extent, on the nonliving substrate, or "parent material," from which it developed. Grassland soils still bear the mark of the glaciers that retreated thousands of years ago. For instance, the region immediately south and west of the Missouri River (from Montana and the Dakotas across to northwestern Nebraska) escaped the direct impact of the ice. As a result, the grasslands there took root on an ancient, unglaciated landscape that had originally been shaped by the shallow seas of the dinosaurian age. Because of the chemical composition of those ancient deposits, the soils that formed on them are not particularly rich in some of the minerals that growing plants need.

By contrast, soils to the north and east of the Missouri (from the Canadian prairies to northeastern Nebraska) have been built on mineral-rich gravels and

ground-up rock left behind by the glaciers. And large parts of the central plains (from southeastern Wyoming, south into Texas and east to Kansas and Missouri) are blanketed in thick drifts of silt, or loess, that were deposited by fierce windstorms around the end of the Ice Age. Whether ice-borne or wind-borne, this glacial debris has provided the soil with a generous resource of plant nutrients that enhance its fertility.

In a number of places across the plains, the Ice Age also left behind a legacy of sand, a material that is much less favorable than silt for soil development. Sand is composed of relatively large particles, with large pores in between, a structure that permits water to run through it freely. As a result, organic material is flushed away before the soil organisms have a chance to set to work and very little humus is produced. In the millennia since the glaciers' last retreat, sand prairies like the Nebraska Sand Hills and the Great Sand Hills of Saskatchewan have produced only a fragile layer of topsoil that is susceptible to every passing footfall or gust of wind.

The active sand dunes south of Monahans, Texas, respond to every ripple of the prevailing winds. Formed from the debris of ancient mountains, the sands were deposited here about 25,000 years ago. Today, they provide habitat for remarkable plants like sand bluestem, sand sage and shinnery oak and for animals like the dunes sagebrush lizard.

Yet the inability of sand to hold onto water also offers an unexpected advantage to certain plants. Because moisture is not tightly bound to the sand granules, it is readily available to the roots and often supports surprisingly lush stands of vegetation. The Nebraska Sand Hills, for example, are home to a distinctive community of grasses and forbs, including specialists such as sand bluestem, sand dropseed, sand reedgrass, and the amazing blowout grass, which pushes its rhizomes out to a distance of 40 feet (12 meters), searching for bare sand in which to establish new plantlets. Similarly, the lemon scurfpea, or lance-leafed psoralea—a delicate-looking plant that is actually tough as nails—relies on a network of long, shallow roots that hold the soil in place and allow it to flourish even on the shifting sands of windswept banks.

> THE TAXONOMY OF SOIL

Over the last several decades, North American soil scientists have devoted themselves to the complexities of soil classification with such enthusiasm that they have produced two completely separate systems. The USDA taxonomy, used in the United States, divides soils into eleven broad categories, or orders. Most grasslands soils are classified as mollisols (literally, "soft soils"), a designation that is further subdivided by climatic conditions. Soils may be udic, ustic, xeric, or aridic as they progress from humid to parched; they step up from cryic to thermic and hyperthermic along the vector from frigid to warm. Other terms describe characteristics such as the texture and chemistry of the soil. Thus, a cool, dry soil

from northern Montana is classified as an aridic cyroll (a dry, cold mollisol), while a warm, moist soil from Texas keys out as a thermic udoll (a warmer, wetter mollisol). Meanwhile, the rich farmlands of west central Kansas emerge from the classification process as fine, smectitic, mesic typic argiustolls, or silty, warm, well-watered mollisols with a heavy subsoil laced with slippery clays. And so it goes, in an ardent outpouring of precisely defined vocabulary.

In Canada, scientists have retained an older system that is based more closely on observable soil properties, including the color and depth of the topsoil and the subsoil, or the A and B horizons. According to this system, prairie soils are not mollisols but chernozems, a term that comes from the Russian for "black earth." But since chernozemic soils actually vary noticeably in color, this order is further subdivided into four major groups: Brown, Dark Brown, Black, and Dark Gray soils. Yet even this apparently straightforward system shows signs of linguistic strain as scientists bend to the task of distinguishing one patch of prairie soil from the next. Is this a Calcareous Brown soil (light colored and high in calcium) or a Gleyed Solonetzic Dark Gray soil (somewhat darker, saline, and poorly drained)? Soil science is clearly not a profession for the tongue-tied.

Profile of a grassland soil, showing topsoil, A; subsoil, B; and parent material, C.

A

B

C

Alfisols
soils with a surface zone of silicate clay
moist to wet, warm soils
dry, warm soils

Aridisols
soils of arid environments with subsoil development
dry, warm soils

Entisols
soils with little morphological development
dry, warm soils
dry, warm, sandy soils

Mollisols
grassland soils
moist to wet, warm soils
dry, warm soils
moist to wet, cool soils

Vertisols
clayey soils with high shrink/swell capacity
warm soils of all soil suborders

Utilisols
acidic soils with clay-rich subsoils
warm soils of all soil suborders

Inceptisols
moist soils with little subsoil development
warm and cool soils

0   500 kilometers

SOILS OF THE GREAT PLAINS

The nature of sandy soils is largely determined by their geological origins. But most soils on the plains are even more strongly influenced by a second physical force—the climate. In particular, soil development is especially susceptible to regional differences in temperature and precipitation. In general, the more moisture that is available to grasses, the more roots they produce and the deeper and darker the topsoil that accumulates. Because precipitation increases along a west-to-east gradient across the Great Plains, soils follow a similar trend, increasing in fertility across the region. The thin, light-colored soils of the western steppe grade into the deeper, richer soils of the mixed grasslands to the north and east and culminate in the thick, black earth that developed under the tall-grass prairies.

Originally hauled to the plains by the glaciers thousands of years ago, this boulder is now being cracked apart by frost wedging and transformed into soil.

Inevitably, in a world of intricate interactions, this simple pattern is complicated by a conflicting variable, namely, a north-to-south gradient in temperature. Where average soil temperatures are low, as they are on the northern plains, the soil organisms are sluggish and the rate of decomposition is slow, so humus accumulates in the topmost horizon. In the warmer south, by contrast, the more active microbes eat around the clock, causing rapid decomposition and slower rates of humus production. A glance at a generalized soil map confirms this effect. The rich black "mollisols" of the northernmost plains gradually give ground farther south to "alfisols" and "aridisols"—thinner, less fertile soils. At the same time, it is also possible to detect an overall improvement in soil quality from west to east, under the influence of increasing precipitation.

Movers and Shakers

If the climate has mediated the development of prairie soils over several thousand years, it also sets the pace of life in the here and now. Although the weather in the underworld is relatively stable compared with the chaos that reigns above ground, the soil ecosystem is not immune to climatic disasters. Drought is a particular challenge. Most soil organisms are aquatic species that have become adapted to live in the film of water surrounding the soil grains. If this habitat dries up (whether for a few weeks in the heat of midsummer or during a prolonged drought), the microbes are thrown into a crisis. Although many individuals die, many more survive by encasing themselves in protective armor, transforming themselves into cysts, and becoming dormant. Waterbears, for example, shrivel up into little kernels called "tuns" that are capable of withstanding unimaginable cold (approaching absolute zero), unspeakable heat (well above boiling), and brutal doses of Xrays (a thousand times the lethal dose for humans). When a piece of moss in a museum collection was moistened after 120 years, the long-dormant waterbears roused from their sleep and immediately started grazing.

Although waterbears may set the outer limits of biological endurance, most soil microorganisms have an extraordinary ability to "play dead" for months or years. A nematode egg can lie in the soil for up to a decade, waiting for its preferred plant host to appear. The minute that conditions improve—for example, when a dry crumb of soil is moistened by long-awaited rain—the soil organisms come back to life and start to eat and reproduce at a hectic rate. Life occurs in manic pulses—wet, on; dry, off—typically with peaks in the spring and autumn, and lulls in the hot summer months.

But there is one major group of soil organisms that has not adapted comfortably to the erratic moisture regime of the Great Plains. Earthworms, those most celebrated of all soil dwellers, breathe through their moistened skin, and for them, desiccation can be fatal. While most species can survive moderate drying by burrowing deep into the subsoil, coiling up in a knot, and entering dormancy, few can survive severe or prolonged water shortage. In the whole of the Great Plains, earthworms are naturally abundant only in the relatively well-watered soils of the east and southeast, including all or part of Nebraska, Iowa, Kansas, Oklahoma, Texas, and Missouri. In this region, the soil is home to worms of two dozen different species—gray, pink, brown, massive or delicate. Some are twice the length of a man's outstretched hand; others would fit neatly on a fingertip.

Among the most common earthworms in these soils are the so-called peregrines—hardy European and Asian species that have been transported to the prairies (and around the world) in the roots of potted plants. Ideally suited for life in agricultural soils, they have expanded their range tremendously in the last few hundred years, often at the expense of indigenous species. Of the six families of earthworms native to North America, four are known to occur naturally in tall-grass prairie soils. The populations currently seem to be strongest in relatively untouched regions such as the Flint Hills of Kansas, though even there the peregrines continue to intrude.

Native earthworms: genus Lumbricus, *left*; Diplocardia, *right*.

Nobody really knows what difference it makes when foreign earthworms appear on the scene, but it must have a profound effect on soil ecology. Earthworms are big eaters, and where they are abundant, they serve (in Aristotle's undying phrase) as "the intestines of the Earth," capable of producing their own weight in nutrient-rich castings every twenty-four hours. But not all earthworms are created equal, and experiments hint that native species may be specially attuned to meeting the nutritional requirements of native plants. When big bluestem was grown in pots with either the native earthworm *Diplocardia smithii* or the introduced pasture worm *Aporrectodea turgida*, the grasses produced significantly more roots in the presence of the homegrown species. Perhaps *D. smithii* squirmed in closer to the roots and delivered the nutrients where they were needed most. Or maybe the native worm produced hormonelike "growth factors" (as earthworms are known to do) that were particularly effective in stimulating the tissues of its longtime evolutionary companion.

In one way or another, the introduction of alien life-forms into tall-grass prairie soils must certainly be having complex repercussions. And the disturbance is presumably even greater in the northern and western plains, where earthworms appear to have been rare, or entirely absent, before settlement began. Across the entire northern span of the continent—to the southern limit of the most recent glacial advance—there are essentially no native species of earthworms. Either such creatures never existed or they all perished under the ice, never to be seen in modern times. (One telling exception is a small, pallid species called *Aporrectodea bowcrowensis* that was recently discovered in Alberta's Porcupine Hills, in an area that was not glaciated.) Native species of earthworms are also virtually absent from the short- and mixed-grass prairies of the western and central plains, a region that is chronically short of rain. The only earthworms to occur there are the immigrants that arrived with settlement.

> DUNG BEETLES

Death by desiccation is a threat to many insects, particularly to the soft-bodied, unprotected larvae. One way to avoid this hazard is to take shelter underground, away from the drying effects of the wind and sun. This need for protection from the elements explains why so many fly and beetle larvae—maggots, wireworms, and grubs—are found in the soil, within wriggling reach of subsurface moisture.

As a rule, the adult forms of these insects are not attentive parents: they simply lay their eggs in the soil and leave the larvae to look after themselves, as they grow, pupate, and eventually take wing as a new generation. Not so the members of the family Scarabaeidae, or dung beetles. Despite what may seem a lowly station in life as eaters of excrement, dung beetles are sophisticated insects that invest in parental care by providing their offspring with food, moisture, and shelter. On the Great Plains, the 100 or so species of dung beetles accomplish this task in one of four ways. First, there are the "rollers," largish beetles (up to an inch, or 3 centimeters, in length) that cut chunks of moist dung from a bison or cow pat and use their forelegs to form it into a ball. Then, engaging their hindlegs and walking backwards, they shove the ball through the grass until they find a suitable place to bury it. In this underground nursery, the female lays an egg inside the ball of dung, which will provide sustenance and protection for the larva.

The second group, the "tunnelers," follow a similar strategy. But instead of struggling to transport the dung cross-country, these species (which tend to be smaller than the rollers) simply chew out chunks of dung and dig them directly down in the dirt, either under or beside the dung pat. Here again, the female lays her single egg inside the lump of dung, which will be the larva's only source of sustenance during its two-to-three-week period of metamorphosis.

Then there are the "dwellers"—the smallest of the lot—which take an even simpler and less laborious approach. Rather than carrying dung down into the

Cow pat with fungus
Candace Savage photo

soil, they lay their eggs right inside the pat. Because dung pats dry out quickly, these insects have a short reproductive cycle, which speeds them from eggs to adults in a few days' time.

Finally, there is a unique group of pat-dwelling beetles that, over the last ten thousand years, have become adapted to eat rodent dung and to live in the cool, moist environment of rodent burrows. Some of these species are at home in open-mouthed burrows, like those made by ground squirrels and black-tailed prairie dogs, while others are found exclusively in the closed burrows of pocket gophers. On the rare occasions in spring and fall when the pocket gophers open up their dens, the air is sometimes peppered with thousands of tiny dung beetles making their escape, as they search for new burrows and fresh opportunities.

Bite by bite, dung beetles and dung-beetle larvae process enormous amounts of organic matter and make an incalculable contribution to the fertility of the soil.

Although little is known about their progress on this new frontier, they appear to be expanding into both grasslands and croplands, boldly going where no worm has gone in at least 10,000 years.

So, despite everything we've heard to the contrary, the presence of earthworms in the soil is not always a mark of good health. Across much of the Great Plains, it is a prime indicator of disturbance. In arid regions, the work of digesting organic matter is assigned by nature not to tender worms but to the invincible legions of drought-hardy microorganisms. Meanwhile, an unlikely consortium of animals has taken on some of the earthworm's other tasks. For instance, the job of turning the soil—digging in dead vegetation from the surface and bringing up clay from below (a service that is provided by earthworms where they are plentiful)—is here performed by burrowing mammals, including ground squirrels, pocket gophers, prairie dogs, and badgers, and most notably by ants. Given their impact on prairie soil, it could almost be said that ants are the earthworms of the Great Plains grasslands.

There are dozens of different species of ants on the prairies (eighty-three, for example, in North Dakota alone), and while they differ in many particulars—size, color, diet, habitat—they tend to have one characteristic in common. Unlike ant species elsewhere that may take shelter in rotten logs or inside rolled-up leaves, most of the ants on the grasslands nest in the soil. An anthill is merely the visible extension of a complex of tunnels that extends down through the topsoil and into the subsoil beneath, often to a depth of 3 to 6 feet (1 to 2 meters). The labor of excavating these underground runways and lugging the soil up, grain by grain, can occupy a colony of several thousand ants for the better part of a decade.

But this Herculean effort is justified by the benefits that accrue to the ants. Like all insects, ants are ectothermic, or cold-blooded. In other words, they assume the temperature of their surroundings and regulate their body heat by seeking shade or sunlight, as required. By constructing their complex, multilevel dwellings, ants alter their environment to create a range of microhabitats, each with its own particular microclimate. In winter, for example, they may retreat to their deepest chambers (below the reach of the frost), where they shuffle groggily through the coldest months. In summer, some species make sophisticated use of their anthills, which are carefully angled to soak up rays from the sun. As heat builds up in the hill, crews of workers lug their larval brood from place to place inside, striving to keep both themselves and their charges at just the right temperature.

facing page: The Nuttall's cottontail lives in brushy ravines and rocky breaks across the western plains. In spring, the female rears her young in a cozy, underground nest, often appropriating a burrow dug by a prairie dog or badger.

Formica inserta

> AN ASTONISHMENT OF ANTS

The fascination of ants extends far beyond their importance in building the soil. Although prairie ants represent only a tiny fraction of the 15,000 species that have been described around the world, they nonetheless exhibit a full range of weird and wonderful behaviors.

By and large, prairie ants are carnivores with a taste for sweets. Because they hunt in groups, they are generally the most successful predators in their size range, subduing impressive numbers of spiders and smooth- and soft-bodied insects. Prey animals are generally torn into bits and carried back to the nest, where they are fed to the larvae. The worker ants' main source of nourishment, however, is nectar from plants. This they typically obtain from one of two sources. Many plants, including native sunflowers and partridge peas, have special glands on their leaves, stems, or buds that produce ant-sized drops of nectar. Ants come to sup at these honeyed fountains—or "extrafloral nectaries," as they are formally known—and defend them vigorously, in the process inadvertently defending their hosts against plant-eating insects.

Ants also acquire nectar secondhand, by milking "honeydew" from the anuses of sap-eating aphids. Some ants carry on this trade in public, on the leaves of aphid-infested plants, but others prefer to milk their aphids in the privacy of the soil. Except for nuptial flights—when winged "reproductives" burst forth to mate and form new colonies—these ants spend their entire life cycle underground. Their nests feature special chambers around the roots of selected plants, where they keep their "herds" of root-feeding aphids. The ants tend their livestock constantly, tapping the aphids with their antennae to induce them to secrete sweets and slaughtering the tender young to supply their need for meat.

Lasius neoniger

facing page: The Richardson's ground squirrel, or prairie gopher, is one of a hardworking cadre of burrowing mammals that aerate and turn the soil as they tunnel through it.
Arthur Savage photo

As ants manipulate their surroundings to meet their own needs, they inadvertently replenish the soil. The particles of clay that they carry up from their basements are rich in the minerals needed by plants and thus help to maintain the fertility of the upper layers. One estimate suggests that the top 2 feet (60 centimeters) of prairie soil is turned over every hundred years through the infinite efforts of ants and other soil creatures. And all this is achieved without any attention from us, beyond the farthest horizons of our senses.

five

HOME ON THE RANGE

Oh, give me a home,
Where the buffalo roam,
And the deer and the antelope play,
Where never is heard a discouraging word,
And the sky is not clouded all day.

**HOMESTEADER BREWSTER HIGLEY,
ON THE BANKS OF BEAVER CREEK, KANSAS, 1872**

IF I HAD to name my favorite place on the prairies, it would be the high, arid benchlands that sweep along the rim of the Frenchman Valley in southern Saskatchewan. Climb up there on a blue day in early September, out and onto the bald steppe at the top, and the wind will slam against you as if it had a grudge against anything that dared to raise its head above the grasses. This is a landscape that has held to the horizontal for thousands of years, even resisting the torrents of meltwater that once rushed down from glaciers in the Cypress Hills and gouged out the wide, flat trench of the river valley. From up on the benches, you can see the descendant of this ancient flood, a soapy, sleepy little stream that writhes through its oversized course, as if trying to make up in complexity for what it has lost in force.

High on the benchlands, the grasses are stunted and crisp, and the ground bristles with clumps of prickly pear cactus. At one time, long before the Ice Age, this drought-stricken upland was itself the course of a great river that flowed down from the young Rocky Mountains across the northern plains, burdened with loads of gravel and debris. As stones were dropped

When a Kiowa woman named Old Lady Horse looked back on the past, she recalled the not-so-distant time when all the necessities of life had been provided to her people by the bison. Hides for shelter and clothing, bones for tools, blood and meat for food. "The buffalo were the life of the Kiowa," she said.

When Europeans came to the plains to build railroads and raise cattle, the bison did their best to protect the Kiowa from harm. "They tore up railroad tracks and the gardens," Old Lady Horse recollected. "They chased the cattle off the ranges. The buffalo loved their people as much as the Kiowas loved them."

But when the newcomers sent in soldiers and hide hunters, the buffalo admitted defeat and gathered in council to decide what to do. As it happened, the Kiowa were camped on the north side of Mount Scott, Oklahoma, at this time. Early one morning, a young Kiowa woman looked up from camp toward the mountain, through the mists over Medicine Creek, and saw the last bison herd appear like a spirit dream. As she watched, the face of the mountain opened and the bison walked inside, into a world of plum blossoms and freshness, where "the rivers ran clear, not red." Into this world of beauty the buffalo walked, and the mountain closed behind them and they were gone.

Old Lady Horse told this story to Alice Marriott, who included it in her book *American Indian Mythology* in 1968.

by the river, they settled onto the riverbed, where they formed a thick layer of sediment. (As the ages passed, this pavement helped the land resist erosion by water and wind, with the result that the former valley is now a high tableland.) Millions of years later, the benches of the Frenchman Valley are still strewn with rounded, river-washed stones, most of them two or three times the size of a clenched fist, all speckled and splotched with lichens.

The eye scans the stone-pocked surface, searching for clues. Grass, sky, rocks. Scattered rocks that suddenly swerve to follow a curve, a curve that coalesces into a circle. Two concentric circles of stones lie half-buried in the grass, tracing a circumference of about twenty paces. A few steps away, another ring and another. People have been here before us in the fairly recent past, though whether it was two hundred years ago or two thousand is more than we can tell. All we know is that a party of bison hunters once chose this bank of the valley to make their camp and that they used these rings of rounded stones to anchor their skin tents. Downslope from the tipi rings, where the benchland begins to fold toward the valley, smaller rocks and pebbles have been washed into terraced drifts, and some of them are chipped and sharpened along one edge as if they had been fashioned for working skins. They fit neatly into the hand.

This land has never been disturbed by the plow—not even a stone has been touched—and it would be easy to think of it as wild, the last stand of the great

North American grasslands. In the beat of the wind, you can almost believe you are hearing the muffled drumming of a bison herd that, any minute now, will come rolling into view over the horizon. Through bones and stones and life-forms, these lands conserve not only the memory of the past but the whole promise of a future for many grassland animals and plants. The surviving native grasslands bring us as close as we can now get to the prairie in its natural abundance. Yet when you walk toward that horizon and peer down into the valley beyond, you will not find wild herds or camps of nomadic hunters. Instead, you are very likely to see a bunch of cows. It's still fabulous out there on the rangelands, but it ain't exactly wild.

The Intercontinental Bovid Boogie

Rangelands—expanses of native grassland that are grazed by livestock—exist only where the prairie has somehow managed to escape the plow, usually because the soil is too dry, too thin, too rocky, or too steep to be suitable for crops. The greater the agricultural potential of a region, the less native prairie is left. Because moisture and soil fertility improve from west to east and from south to north, the percentage of the land in natural cover increases in reverse, from east to west and from north to south. Out west on the short grasslands, where the rains are meager and the soil is relatively poor, something on the order of 70 percent of the landscape has survived with its natural integument of grasses and wildflowers. A little to the north and east, by contrast, in the slightly moister mixed-grass zone, the percentages drop off, and what little prairie survives is badly fragmented. (In Manitoba, to cite the most extreme example, less than 0.1 percent of the original mixed-grass landscape remains intact.) On the whole, the irresistibly fertile tall-grass zone has fared worst of all, with less than 1 percent of the entire ecoregion remaining in a more-or-less natural condition. This state of affairs earns the tall-grass prairie dubious honors as the most endangered ecoregion in North America. Today, the only large, continuous blocks of tall grasslands lie in the Flint Hills of eastern Kansas and the Osage Hills of northeastern Oklahoma, where a stony rime of crystalline quartz just beneath the surface long ago put a dent in the enthusiasm of plowmen.

The surviving native grasslands span the complete spectrum of prairie types, from the stony benchlands of the Frenchman Valley and the rollicking dunes of the Nebraska Sand Hills to the sculpted badlands of the Palo Duro and beyond. Ecologically, they are as different from one another as big bluestem is

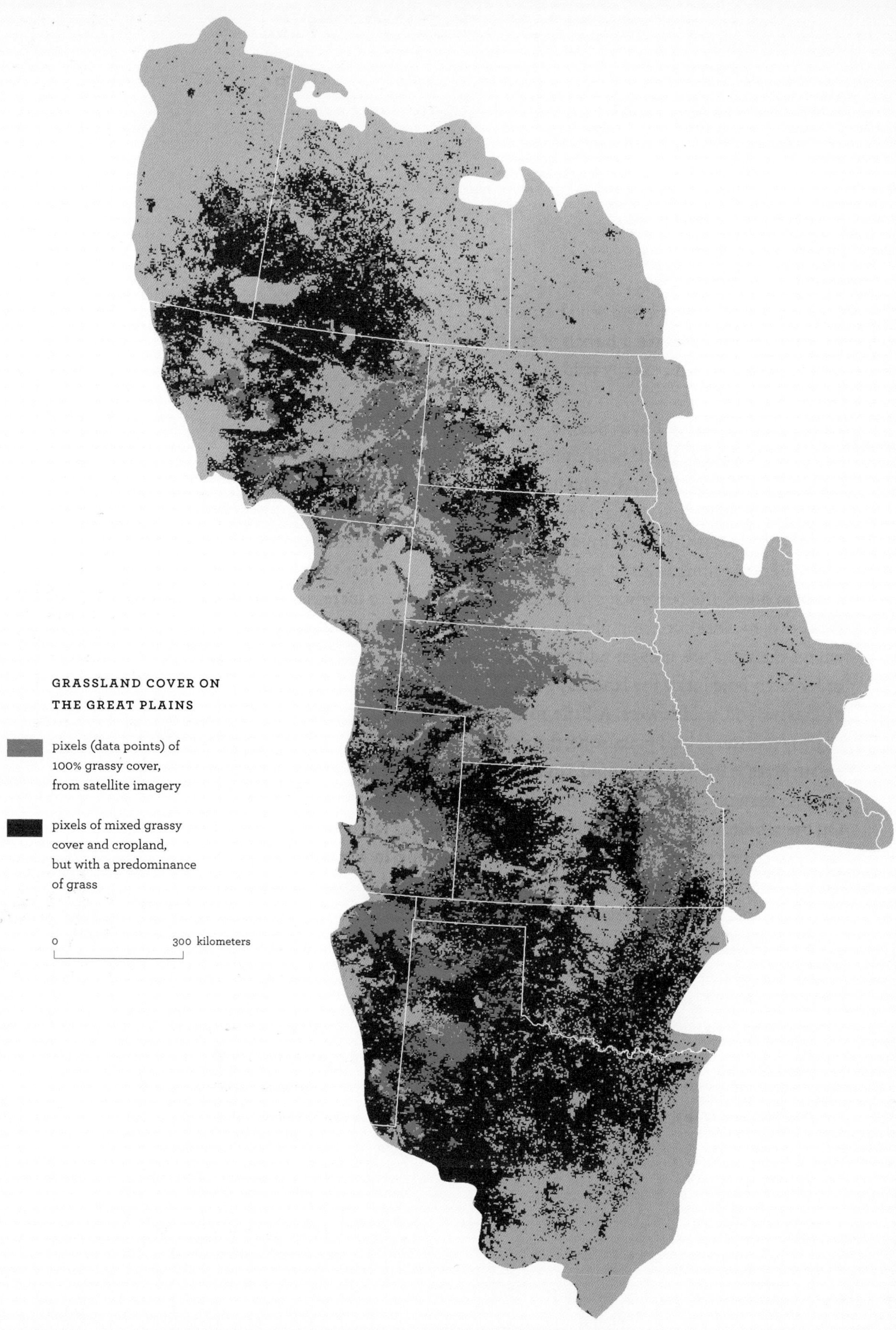

GRASSLAND COVER ON
THE GREAT PLAINS

pixels (data points) of
100% grassy cover,
from satellite imagery

pixels of mixed grassy
cover and cropland,
but with a predominance
of grass

0   300 kilometers

from blue grama or as a sage grouse (a threatened bird of arid prairie) is from a prairie chicken (its threatened cousin on the humid tall grasses). Yet for the last century or so, these diverse grasslands have all answered in similar ways to one set of human demands—to produce food for large numbers of cattle and ultimately for ourselves.

In some ways, the introduction of domesticated livestock onto the Great Plains was not much of a shock to the ecosystem. Bison and cattle belong to the same evolutionary lineage, the family Bovidae, and trace their ancestry back to India and China some 2 million years ago. But whereas the ancestral bison headed north across the Bering land bridge into the Americas, cattle took a turn to the west and wandered through the Near East and into Europe. (They are thought to have been domesticated in or around what is now Turkey some 10,000 years ago.) True to their common ancestry, the two species filled much the same ecological role on their respective continents, as large mammalian herbivores that specialize in consuming grasses and, to a lesser extent, other forbs, or nonwoody plants.

Since grasses are ribbed with particles of silica (grains of glassy sand), both bison and cattle have self-renewing teeth that grow in to compensate for the wear and tear. And since grasses and other plants are composed mostly of cellulose, which mammals cannot digest, both species have fermentation vats

His body a blur of intensity, a male sharp-tailed grouse performs his mating dance. In spring and fall, groups of male grouse meet at leks, or dancing grounds, to compete for the attention of the females.

Greater prairie chicken

PRAIRIE
CHICKEN
DISTRIBUTION

■ Greater prairie chicken
■ Lesser prairie chicken

0   500 kilometers

Sharp-tailed grouse
Greater sage grouse

0   500 kilometers

GROUSE DISTRIBUTION

> HOOT, BOOM, STRUT, DANCE!

Four species of grouse are native to the Great Plains: the sharp-tailed and greater sage grouse, and the greater and lesser prairie chickens. Despite the differences in their common names, they are so closely related that, where their ranges overlap, they occasionally interbreed and produce fertile hybrids.

Historically, the areas of overlap were significant. Greater sage grouse, which, as their name suggests, prefer the dry, sage-dominated grasslands of the far west, nonetheless formerly occurred as far east as Nebraska. Prairie chickens, by contrast, were most common in the lush tall- and mixed-grass prairies of the eastern plains, but their range extended west to Alberta and Texas. (Of the two species of prairie chickens, the greater is found from the central plains north, while the lesser has a more southerly distribution.) The sharp-tail held the middle ground, occupying a variety of short-grass, mixed-grass, and grass-shrub habitats from central Canada to Oklahoma.

Although sharp-tails are still common throughout much of their range, their southernmost populations have vanished in the last hundred years, the combined result of overhunting and habitat loss. Sage grouse and prairie chickens persist as scattered and declining populations on fragmented habitat, and their long-term survival is in question. But who could bear to lose them, if only because of their antic, exotic, ecstatic performances? In the cold dawn of early spring mornings, the male grouse gather (each species in its own group) to compete with each other for status and access to females. First, they establish themselves on an open piece of prairie (stubble or closely grazed grass), where each bird stakes out a miniature territory for himself. Within this ritual domain, he faces off against his neighbors in a twirling, gobbling dance, puffing up pouches on his neck or chest to maximum effect. The object of these strange proceedings is to claim a territory at the center of the lek and thereby impress the hens as a top-quality prospect. The dominant male in the lek fathers most of the chicks and makes his contribution to the survival of the species.

for stomachs—elaborate four-chamber structures in which a frothing stew of microorganisms breaks down the tough plant fibers. Halfway through this process (between stomachs 2 and 3), the animals cough up a "cud" of half-digested mush, give it a thoughtful second chew, and send it back down for another round of microbial demolition. By collaborating with these small-but-mighty microbes, large ruminants like cattle and bison free up nutrients locked inside plants and keep resources cycling through the ecosystem.

As they convert plant tissues into cold-cuts-on-the-hoof, herbivores inadvertently provide a source of highly digestible food for the next trophic level in the system, the carnivores. The carnivores, in their turn, inadvertently assist the herbivores in avoiding predation. By removing the halt, lame, and unwary from the breeding stock, predators select for individuals that are best adapted to escape from their grasp and hence make successive generations of their prey more difficult to catch. Thanks to this rigorous breeding program, bison and cattle have both acquired large eyes and excellent vision, the better to help them detect predators. They have become strong and sure-footed, able to sprint away from danger or to travel long distances to seek safety. Most important of all, both species have become overwhelmingly sociable. Like many other animals that live on the plains, bison and cattle find comfort in the company of others of their species. Not only can they benefit from the combined vigilance of the group—all eyes peeled for a possible threat—but they can also attempt to use their neighbors as "bovine shields." The more potential victims that are available to a predator, the less chance there is that any given individual will end up as a sacrifice to trophic exchange.

So even though they spent the last half million years or so on opposite sides of the Atlantic, bison and cattle are fundamentally alike. Removing wild American bison and replacing them with tame Eurasian cattle—though a stunning act of hubris—was ecologically relatively neutral, the substitution of one large, ruminant herbivore for another on a landscape that had sustained large herds of grazers since the retreat of the glaciers. (The same cannot be said for the desert and semidesert grasslands west of the mountains, which had not evolved under heavy grazing pressure from bison and which have suffered serious damage since livestock were introduced.) The ease of this transition is demonstrated by the suite of animals that once lived in, under, and around bison pats. As nearly as anyone can tell, this group, which includes several dozen species of dung beetles and several specialized mites among many others, took to their new diet of cow pies without hesitation. And the species that dine on the dung dwellers—a diverse group that includes blackbirds, small rodents, and (in the tall-grass prairie) the beautiful little ornate box turtle—also seem to have made the switch without any noticeable problems.

How Grazing Shapes Grasslands

This is not to say that the introduction of cattle to the Great Plains has been completely benign. It has not. Cattle and bison, though similar, are not identical. A case in point: bison like to throw themselves on the ground and flail around in the dirt, a self-care routine that is thought to coat the skin with dust and offer protection from biting insects. In the process, they wear away shallow bowls, or "wallows," in the earth. By rubbing out the grasses from these hollows, bison create openings for other kinds of plants, often moisture-loving species such as prairie dropseed, together with various rushes and sedges. Because they increase the diversity of plants that are available for shelter and food, bison wallows also augment the diversity of insects, birds, and mammals that the local landscape can support. And if the depressions fill with water (as they may do in the spring), they provide seasonal habitat for aquatic insects and water-loving shorebirds. Or at least they used to.

Ornate box turtle

Because cattle do not wallow, this dynamic has been lost. Meanwhile, cattle have pushed the ecosystem in new directions. Less well adapted than bison to extremes of temperature and drought, cows spend more of their time in the shelter of trees and around sloughs and watercourses. This behavior severely affects both water quality and the condition of riparian, or bankside, forests. What's more, although cattle and bison are both grazers with a special fondness for grass, there are subtle but significant differences in their diets. Bison prefer a steady diet of grasses, with just a garnish of other plants (forbs and shrubs make up only 5 to 10 percent of their diet). Cattle, by contrast, choose grass as their staple food but they also enjoy a side salad of mixed greens (forbs and shrubs provide between 20 and 40 percent of their feed). Although no one knows how the vegetation of the rangelands has changed since the bison were displaced—which grasses and forbs have become more common and which more scarce—Bossy the Cow's exotic dietary preferences must have had an impact.

Herbivory shapes grasslands. The first principle of range management is that grazing causes shifts in the relative abundance of the many and various species of grassland plants. In mixed-grass prairie, for example, blue grama, buffalo grass, and pasture sage tend to become more common on range that is grazed by cows; these plants fall into the class known as increasers. Conversely, midheight grasses such as needle-and-thread and western wheatgrass become more scarce and hence are called decreasers. (It has even been suggested that the short-grass prairie, with its mat of increasers like blue grama and buffalo grass, is really just mixed-grass prairie that has been subjected to heavy grazing.) Some plants—like the big, bold butterfly milkweed of the tall-grass prairies—vanish from the range the second they appear because the cows enjoy eating them. Out in the pasture, grizzled rangemen shove their Stetsons back off their brows and lean against their pickups to discuss the status of these "ice-cream plants" in their pastures.

Until fairly recently, range managers have worked on the comforting assumption that the disruptions caused by grazing could be corrected simply by removing the cattle and letting the pasture rest. This belief was based on what seemed at the time to be a sure basis in science: the cogent theories of a pioneering ecologist from the University of Nebraska, Frederic E. Clements. As Clements explained it in the 19-teens and 1920s, mature, natural grasslands were the endpoint, or climax, of a linear process of plant succession. This climax

By rolling, or wallowing, in the dust, bison wear hollows in the earth, creating seedbeds for plants and catch basins for water.

Arthur Savage photo

Butterfly milkweed flourishes along roadsides in the Flint Hills of Kansas but is conspicuously absent from adjacent pastures, where cattle can get at it. Like other species of milkweed, this one is also a favorite food plant of butterflies, notably monarchs.

facing page: Unlike most of the wild flowers on the plains, the prairie sunflower is an annual rather than a perennial. It is often one of the first species to colonize a site following a disturbance.

community, he said, represented a state of equilibrium in which the conflicting needs of the various plant species (their demands for nutrients, water, and light) were held in such fine balance that the grasslands, if left to themselves, would wave on unchanged forever. But any disturbance, such as heavy grazing, fire, or drought, upset this steady state and sent the system tumbling back into an immature and unstable stage. Cattle, for example, not only evicted the ice-cream plants and decreasers from the climax community but in so doing opened the prairie to early successional species such as broomweed and Russian thistle. These same "weedy," or invasive, plants were always first on the scene when plowed lands were abandoned or pastures were degraded by overgrazing.

The good news, as Clements saw it, was that plant succession was a predictable and reversible process, in which pioneering species inevitably created conditions for their successors, which created conditions for their successors, and so on, until the climax community was ultimately restored. Thus, any damage caused to the rangelands by overgrazing could be repaired, in due course, through the healing processes of nature. If only this were so. Instead, it turns out that succession is more fickle than the good professor believed—more subject to the vagaries of climate and simple luck. Once the prairie has been disturbed, there is no great likelihood that it will grow back in, plant for plant, just as it was before it was disrupted. It is even possible that it will not come back at all. In the last hundred years, for example, large areas of the southwestern United States have been permanently converted from grassland into a prickly shrubland of mesquite and creosote bush, probably as a result of heavy grazing during the 1800s and 1900s.

Don't Fence Me In

But the main ecological challenge on the rangelands—the real thorn in the side of the ecosystem, so to speak—is not the absence of bison or the presence of cattle. It is barbed-wire fencing. The "devil's rope," as it was once known, went into mass production in the mid-nineteenth century to meet the needs of settlers moving onto the treeless (post-and-rail-less) plains. By the early 1880s,

Leafy spurge hawkmoth

> ### ALIEN INVADERS

Since the mid-1800s, hundreds of plant species have been plucked from their native surroundings, carted halfway around the world, and introduced into the grasslands of North America. Although many of these newcomers have to be coaxed along, others have flourished in this home-away-from-home. For example, Eurasian grasses such as crested wheatgrass, smooth brome, Kentucky bluegrass, and tall fescue were initially planted on restricted acreages to provide forage for cattle or (during the drought of the 1930s) to protect the soil from erosion. Since then, however, these exotic species have advanced beyond all bounds, outcompeting the native species and eventually crowding them out.

Happily, these intrusions are not always a dead loss. To a herbivore looking for fodder or a songbird building a nest, grass is a useful resource, whatever its evolutionary origins. So even though these plants arrived as intruders, many of them have become contributing members of the ecosystem. But there are notable exceptions. Leafy spurge, for example, is thought to have arrived from Europe in 1827, as unspecified cargo in the water ballast of a sailing vessel. Freed from European diseases and insect pests, the newcomer laid claim to the continent, infesting both tilled fields and overgrazed pastures. Even today, leafy spurge continues to advance, especially across the northern rangelands. Not only does it outcompete native plants for nutrients, water, and light, it also exudes toxins from its roots that prevent other plants from putting up shoots. What's more, it is defended against herbivores by poisons in its leaves, which are repellent to cattle, horses, and many wild species. Among the few animals that can eat it are certain species of European flea beetles and caterpillars that have been introduced to limit the plant's spread, but even they cannot eradicate it. Resistant to every known method of control, leafy spurge—and other plants of the same ilk—are a menace to grassland ecosystems.

factories in Illinois were spinning out enough barbed wire every year to make twenty-five complete twists around the equator. Not surprisingly, the Great Plains were quickly crisscrossed with fences, and by the close of the decade, the era of the open range had ended. With it went the free interaction between grazers and grasslands.

Before fences, there had been movement. Every spring for millennia, the bison had poured out of their winter ranges on the foothills and parklands, massed together in vast herds, and flowed across the continent on a quest for greener pastures. Like the barren-ground caribou of the tundra, they had charted new routes each year, jogging east to track recent rainstorms, north to follow the spring, south to savor the green-up after a fire. But wherever their travels took them, they always looped west again in the fall and followed a halting, zigzagging course back toward their wintering grounds. There, in the dead of winter (when the tall grasses of the eastern plains had little value as food), the bison sustained themselves on a steady diet of sun-cured, protein-rich blue grama and buffalo grass. Thus, season by season, the bison tracked the resources of the plains and responded to the interactions of the climate, the soil, and the plants.

What happened wasn't always pretty. Even with the freedom to move at will, tens of millions of bison couldn't always find enough to eat. In fact, the fossil record suggests that the species disappeared from the plains entirely for the better part of 3,000 years, between about 4,500 and 7,500 years ago, when the climate was much drier and warmer than it is at present. More recently, during the ferocious winter in 1876, several million bison starved to death along the Brazos River of southern Texas, so many that their humped carcasses reminded one observer of a field of pumpkins.

But if bison sometimes felt the sting of random misfortune, they could also dish it out. After a herd of bison had trundled across the prairie, the prairie generally looked as if a herd of bison had trundled across it. (People traveling in the wake of a herd often complained that they couldn't find so much as a blade of grass for their horses.) Fortunately, grasses are uniquely equipped to put up with this kind of abuse. Unlike shrubs and forbs, which produce new growth from buds on their branches and stems, grasses shoot up from buds at (or even slightly under) ground level. This means that if a grass is eaten, it can immediately send up fresh leaves, whereas a shrub or a forb may be done for until the following spring. What's more, grasses that have been defoliated often shift

Crested wheatgrass

Bluebunch wheatgrass

hormones and other resources from their roots to their new shoots, so that their rate of photosynthesis increases dramatically. The loss of leaves to a herbivore causes some prairie grasses to grow so exuberantly that they slightly overcompensate for the tissues that have been removed. It probably doesn't hurt that the saliva of some grass-eating mammals contains growth stimulants that encourage the grasses to send out new shoots.

Not all grasses are equally well equipped to cope with the stress of being chewed. As a rule, increasers, a.k.a. grazing-tolerant grasses, have a greater capacity to fight back than do decreasers. For example, crested wheatgrass—an introduced species from Asia that can be recognized by its handsome, fishbone seed heads—puts out replacement leaves two or three times faster than does its native relative, bluebunch wheatgrass. This competitive advantage helps explain why crested wheatgrass has been able to crowd out native plants and establish itself across the plains at their expense.

To get the true measure of the relationship between grasses and grass-eating mammals, it helps to see the natural dynamics that evolved between them. Today, this is only possible in the few places on Earth where migratory herbivores are still free to roam, such as the tropical grasslands of the Serengeti Plains in East Africa and the temperate grasslands of Yellowstone National Park in Wyoming. Despite the obvious differences between these two ecosystems, researchers have discovered a surprising similarity between them. In both places, grasses that are grazed, each in their season, by migratory herds produce far more forage than do grasses that are protected from grazing (43 percent more in Yellowstone, 70 percent more in the Serengeti, according to one report). This bonus far outstrips any "overcompensation" that occurs when animals are confined within fenced pastures. (All the same, the management of cattle in fenced pastures does have its advantages, which we will come to in a moment.)

The unexpected synergy between migratory herbivores and grasses is probably the result of a chain reaction that begins when the leaves are cropped. The grass plant responds to this insult by shedding organic molecules from its roots. These molecules provide food for soil microbes, which go into overdrive processing humus and releasing nutrients in a form that plants can use. Thanks to this input of fertilizer, the plants experience vigorous growth, and the range is superabundantly replenished. But this response only appears to happen if the herbivores are free to migrate and select grasses in the juicy prime of youth. If

the stress of herbivory is delivered in a well-timed pulse, the plants may actually benefit from it.

Did the bison and the prairie grasses once engage in a similar interaction? No one knows for sure. But the loss of this cyclical rhythm may be the price we pay for curbing the wild prairie with barbed-wire fencing.

Promoting Patchiness

On the open range, bison (and cattle, for that matter) had gone wherever they chose and moved on whenever they liked. If they traveled through an area quickly, snatching mouthfuls on the run, the grasses were left standing in tattered clumps. If they lingered, the prairie was reduced to a worn-out, close-cropped lawn, where whatever remained uneaten was trampled or splattered with dung. These impacts occurred haphazardly, tracking the uncertainties of rain, lightning, and fire, to say nothing of the whimsies of the bovine mind. Once a herd had passed through an area, it might return in six months or five

Although sometimes found in dry hay fields, LeConte's sparrow is typically associated with wet meadows, where it nests on or near the ground and scurries through the matted grass like a mouse. This individual had taken up residence on the Manitoba Tall Grass Prairie Preserve near Gardenton, Manitoba.

Bobolink

Grasshopper sparrow

Upland sandpiper

years or never again. As a result, the Great Plains were an ever-shifting mosaic, in which a patch of tall, ungrazed vegetation might stand next to a mouthful that had been cropped to the quick. On a larger scale, broad tracts of land might lie untouched (too dry to attract attention), while the green-up after a sweeping prairie fire attracted large herds of bison. Thus, at their most vibrant, the grasslands were a kind of living crazy quilt, with patches of vegetation of varying sizes in varying states of recovery from grazing, fire, or drought. This is the true climax condition of the Great Plains grasslands, a system that achieves stability by responding constructively to continual challenges.

The patchiness created by grazing and other disturbances provides a diversity of habitats for a diversity of prairie residents. (Under present-day conditions, the rotation of cattle from one fenced pasture to the next can be used to simulate this richness and complexity.) Take birds, for example. Of the 260-odd species of birds that regularly breed on the Great Plains, most are species that moved onto the grasslands from other (mostly forested) parts of the continent at the end of the Ice Age. By and large, these immigrant species still cling to the traditions of their ancestral homelands and nest in trees and shrubs. But a small, select group of species—a couple of dozen in all—prefer the open plains, where they nest in, or under, the forbs and grasses.

Within this Lilliputian jungle, the prairie-adapted species seek out the special conditions that each needs for its nests. Some, like the bobolink—a flashy black-and-white blackbird that enlivens the mixed- and tall-grass prairies of the north-central and southern plains with its bubbling song—require dense and expansive stands of thigh-high flowers and grass. The female typically chooses a site that is shaded from the grilling sun (often at the base of a large forb, such as meadow's rue) but where the canopy is open enough to let light filter through. Here, in the cool, dappled shade, she scratches out a patch of bare earth and constructs her finely lined, cup-shaped basket. By contrast, the grasshopper sparrow—a widely distributed but inconspicuous "little brown job" with a shrill, insectlike buzz—opts for more open habitats, with scattered clumps of tangled, mixed-height plants. Its nest is buried in a thicket of vegetation and domed over with grass, an object lesson in the arts of concealment.

Each species looks at the grasslands from its own unique point of view, seeking patches of vegetation that are the right height, density, and structure to meet its needs. Thus, although the grasshopper sparrow and the maddeningly similar Savannah sparrow both are found in open habitats, the former chooses sites with more patches of bare dirt (for catching insects), the latter those with less. And so it goes. The chestnut-collared longspur—so called for the rich splotch of color on the neck of the breeding male—seeks out closely cropped rangelands with the odd rose bush or tuft of grass for nesting cover. Meanwhile, its drabber but doughtier cousin, the McCown's longspur, selects only the most barren prairie, where it nests in the shade of a prickly pear cactus or a cow pat. And the horned lark—another, quintessential short-grass species, much loved for its sweet, slurred song—prefers patches that have been thoroughly trampled, where it can hunker down in the dirt with little more than a stone or a tuft of grass for shelter.

Trying to remember the particulars of which species needs what and why taxes the memory. But for millennia, the turbulent dynamics of the prairies have provided for all these conflicting needs without any conscious intention or evidence of strain. Habitat simply happened. For all its chaotic uncertainties, the prairie ecosystem generated enough stability to sustain even those species with the most complex and demanding needs. The upland sandpiper, for example—a jaunty member of the shorebird tribe that has adapted to life on the arid plains—requires short vegetation with visible perches while courting, knee-high growth with a moderately open structure for nesting, and closely grazed pastures for feeding and rearing its chicks. What's more, these resources must be provided generously enough to support not only a single pair but a loose colony of breeding birds. The sight of an upland sandpiper perched on a fence post is mute proof that these demands have been met in one way or another for thousands of years.

Today, however, two of the main forces that drive the prairie ecosystem have been muted. Wildfire has been suppressed. (On the tall-grass rangelands of Kansas and Oklahoma, annual burns still occur, but they are generally timed to promote the growth of grass and to reduce the diversity of "weeds" and other flowering plants.) In addition, the migratory ebb and flow of herbivores across the plains has been halted. Fortunately, even when confined by fences, cattle help to maintain patches of vegetation—clipped versus tall, open versus dense—and this effect can be enhanced by implementing an appropriate regime of management. By manipulating variables such as stocking level (number of cattle), season and duration of grazing, and rotations of grazing and rest, ranchers can manage the prairie to provide an array of habitats. The best and wisest land managers do exactly that because they understand that rangelands with a natural diversity of vegetation will outproduce and outlast those that are reduced to homogeneous spans of grass. But even with the most beneficent intentions, it would be a tall order to manage grasslands for commercial cattle production and, at the same time, to maintain the prairie in its free flow of creative change.

There is little doubt that the populations of many grassland birds have declined in the last two hundred years. There is no doubt at all that those declines are continuing today. Data from the Breeding Bird Survey (a standardized count of birds that has been conducted annually since 1966) document recent declines in 32 of 37 grassland species. For many of them—including the bobolink, horned lark, and grasshopper, Savannah, and vesper sparrows, together with other favorite prairie species such as longspurs and meadowlarks—the losses are statistically significant. This finding removes any hope that the downturns are merely random blips on the graph. Overall, grassland birds have suffered steeper, more consistent, and more geographically widespread declines than any other group of birds in Canada or the United States. The causes of the losses are no doubt complex and likely include the continued conversion of native prairies to croplands and housing estates. (The less native grassland, the fewer grassland birds.) But these doleful statistics also help underline our failure, as a society, to reward ranchers for managing their land for "patchiness," or biodiversity, and to pay them for producing bobolinks as well as T-bone steaks.

Dickcissel

Food for Sex

Although cattle are the most conspicuous herbivores on the range, they are certainly not alone. Ounce for ounce or gram for gram, natural grasslands support a greater live weight of plant eaters than any other terrestrial ecosystem. (And this statement does not take into account the zillions of nematodes and other soil animals that dine on the plants from below.) The largest group of grassland herbivores, in numbers if not in mass, is insects. While a complete catalog of prairie insects has not yet been compiled, the list of the herbivorous species alone would certainly contain several thousand entries, perhaps even tens of thousands. There are insects that feed on plants by piercing through the cell walls and sucking juices out (aphids and "true bugs"); others that rasp a hole in the plant to obtain nourishing fluids (thrips); still others that are equipped to feed on nectar by lapping (bees), sponging (certain flies),

> THE ROCKY MOUNTAIN LOCUST

Before the introduction of cattle, two very different herbivores dominated the Great Plains. One—the bison—is well known, but the other remains something of an enigma. A grasshopper in everything but name, the Rocky Mountain locust once made its permanent home on the sparse, dry grasslands of the west-central plains (from Alberta and Saskatchewan south to Colorado), where it fed on a mixed diet of buffalo grass and other native plants. Most of the time, it looked and acted like a typical hopper, with relatively short wings and little inclination to travel. But under certain conditions—for example, when the population was stressed by drought—a different breed, or phase, of the species began to hatch, one with longer wings and an unstoppable urge to migrate. Soon, silvery swarms of insects were moving east on the prevailing winds, rising to heights of 6,000 to 10,000 feet (2,000 to 3,000 meters) and sweeping across large distances in a matter of hours. (Grasshopper Glacier, in the Beartooth Mountains of Montana, contains the remains of several million locusts that became embedded in the ice sometime in the distant past, probably when one of these migrating swarms was brought down by turbulence.)

Because the insects could not sustain themselves for long away from their usual range, these eruptions generally ended within a year or two at the most. But for the farmers who moved onto the plains in the 1870s, the year of the locusts could seem very long. The insects fell on gardens and crops like a Biblical plague, eating fruits, vegetables, grains, even the bark off the trees. Then, in the 1880s, the outbreaks inexplicably became less frequent and less widespread. By the early 1900s, the Rocky Mountain locust—a species once so plentiful that it had darkened the skies—had completely vanished.

No one knows why this happened, though the best guess is that cropping, haying, and trampling by cattle disrupted the moist swales and stream-edge habitats that the females needed for laying their eggs. But although the Rocky Mountain locust, *Melanoplus spretus,* is extinct, its lineage lives on in the lesser migratory grasshopper, *M. sanguinipes.* The swarming form of this species can sometimes still be seen high in the summer sky, as a vague shimmering blaze of silvered light.

or siphoning it up with their tubular tongues (butterflies and moths). Several groups have powerful jaws that permit them to gnaw on seeds, like the worker caste of harvester ants, or to feed on leaves and stems, like the hungry legions of sawflies, beetles, grasshoppers, and caterpillars. With their stabbing beaks, hairy tongues, jutting mandibles, and other specialized mouthparts, these insects look as strange, to human eyes, as the weirdest creations of science fiction.

Yet despite their otherworldly appearance, plant-eating insects are of down-to-earth importance in grassland ecosystems. On the one hand, like all herbivores, they prune and shape the vegetation, one miniature mouthful at a time, thereby contributing to the patchiness of the prairie. On the other hand, they

Northern
grasshopper mouse

facing page:
A painted lady
butterfly visits blue
vervain flowers
at the Konza Prairie
Research Natural
Area near Manhattan,
Kansas.

inadvertently provide a generous source of convenient, bite-sized meals for a wide array of small-to-medium-sized meat eaters, including predacious beetles, ants, spiders, songbirds (even seed eaters start their young on an all-arthropod diet), kestrels and other small raptors, and a few oddball rodents—like the thirteen-lined ground squirrel and the northern grasshopper mouse—that are highly insectivorous. But the critical ecological function of plant-eating insects is also their most riveting. They often serve their food plants as sexual go-betweens.

To produce seeds, flowering plants must somehow transfer pollen from the male organ, or anther, to the top of the female organ, or stigma. While some flowers (including the closed, late-season blossoms of wild violets and narrow-leaved puccoons) can accept their own pollen and are thus self-pollinating, most must receive sperm cells from other members of their species. Unable to go a-courtin', some of these cross-pollinated plants rely on the wind to distribute pollen for them. Self-pollination and wind-pollination are both common among grasses. Most wildflowers, by contrast, rely on insects (including beetles, beeflies, butterflies, moths, wasps, and bees) to carry their pollen from plant to plant in a transaction that could be summarized as food for sex. Insects that are enticed into flowers for a feed of nectar pick up sticky pollen grains, some of which they deposit in the next blossom that draws them in for a taste.

Although many nectar-and-pollen-feeding insects are generalists that visit a variety of different plants, some have very specific preferences. For example, of the 200 species of pollinating bees that are found on Missouri prairies, most are widely distributed in other habitats. But a significant proportion of the total—about 25 percent—occur only on native grasslands. Their allegiances are so specific that of the prairie specialists, 24 species work only the uplands, 13 species are restricted to sand prairies, and another 10 species are found exclusively on bottomlands. Different bees for different plants. Sometimes these alliances are easy to observe. In Missouri, for example, the handsome black-and-yellow bumblebee *Bombus nevadensis* is seen almost exclusively on the heads of wild bergamot, where it sends its long tongue into the equally long tubular corollas of the blossoms. By contrast, the all-yellow *Bombus griseocollis*, though more of a generalist, tends to favor plants like purple coneflower and other species of *Echinacea* that hold their nectar within reach of its midlength proboscis. A third species, *Bombus impatiens*, is attracted to the delicate fringe of blooms along the spike of Culver's root, which can be penetrated by its short

tongue. These teams of plants and insects have evolved together to meet the needs of both, providing the insects with a reliable (and sometimes almost exclusive) source of food, and the plants with an eager crew of sex workers. Through this symmetry of behavior and body parts, they have kept the grasslands buzzing and blooming for millennia.

A Prairie Dog Home Companion

After insects, the second-most populous group of herbivores on the grasslands is rodents. Although all the members of this tribe are equipped for a life of gnawing with a pair of big, self-replacing front teeth, different species have become specialized to gnaw on different types of plants. Pocket gophers, for example, are stubby, squirrel-sized mammals with small, weak eyes, that spend most of their lives beneath the surface of the soil. There, they subsist mainly on the roots of perennial forbs, such as yarrow and goldenrod, which they locate by tunneling through the ground. As they "swim" through the dirt, they first kick the soil behind them, then turn around and, using their flat foreheads, shove it back along the burrow and onto the surface. These mounds of loose soil, as tousled as unmade beds, provide openings for annual plants, like bluebur and small lupine, and warm, moist daytime hideaways for tiger salamanders and shiny, low-slung lizards called prairie skinks.

The "pocket" in the pocket gopher's name refers to the fur-lined pouches on either side of the face, in which the animal carries loads of cut-up stems and roots. The same characteristics are repeated in miniature by the pocket gopher's relatives, the pocket mice. These dainty creatures (tail and all, they'd fit into the palm of your hand) also excavate shallow tunnel systems and seal themselves in with dirt, but unlike pocket gophers, they venture out under cover of night to search for food. Their primary resource is seeds, which the prairie showers on them like the proverbial rain of gold. A stand of little bluestem, for example, will typically produce roughly 200 pounds per acre (about 200 kilograms of seed per hectare). To harvest this bounty, the sleek olive-backed pocket mouse (typically found on thin, sandy grasslands across the northern plains) sits up on its hind legs and sifts seeds out of the dirt, then stuffs them into its pouches with its forepaws. Weeds like Russian thistle and pigweed and native plants like butterfly weed and June grass are among its favorites. Seeds that aren't eaten immediately are stockpiled underground. In the fall, the animals move their caches below the frost line and hole up, drowsy but still conscious, for as

Plains pocket gopher

long as six months. During all this time, they do not have to drink because they can extract water from their diet of dry seeds. Ord's kangaroo rat, a bigger, bouncier, desert-adapted species found in the Great Sand Hills of Saskatchewan and other ultradry grasslands, shares this metabolic talent.

But of all the rangeland rodents—pocket mice, harvest mice, jumping mice, pygmy mice, ground squirrels, voles, and more—none is more remarkable than the black-tailed prairie dog of the western plains. Before the prairies were settled, this was the prairie dog's world. Its population ran into the billions, and if all the colonies, or towns, had somehow been amalgamated into one great metropolis, they would have occupied an area as big as Texas. In reality, prairie dog towns were, and in places still are, dotted across the short- and mixed-grass prairies, moonscapes of closely cropped grass marked by craters of bare earth. Each of these earthworks marks the entrance to a shaft that plunges 6 to 10 feet (2 to 3 meters) below the surface and then levels out to lead to a complex of nest chambers and latrines. Prairie dogs descend into their burrows to sleep, give birth, nurse their young, and escape from predators, but they spend much of their lives on their earthworks—sprawled out on their bellies in the sun or upright and alert, ready to bark a warning at the first sign of danger.

Ord's kangaroo rat

On a warm summer day, a prairie dog colony is alive with the hustle and bustle of urban life, as youngsters chase each other, neighbors meet and "kiss," heads pop in and out of holes, and individuals waddle away from their mounds to graze on grass and forbs. But this apparently free-wheeling sociability masks a complex social order based on kinship. Each prairie dog lives in a small, tightknit family group, or coterie, that usually consists of one adult male, three or four breeding females, and nonbreeding youngsters. Young males

No fewer than six species of ground squirrels are found on the rangelands of the Great Plains, all members of the genus *Spermophilus,* or "seed-lovers." Of these, three can be identified by their handsome spotted coats. The sprightly thirteen-lined ground squirrel—which really does have thirteen stripes, plus an elegant grid of spots—is found across the entire span of the prairies and is the most common species in parts of the Dakotas and Nebraska. The Mexican ground squirrel—which sports nine parallel ranks of squarish spots but no dark stripes—is restricted to Texas, New Mexico, and other points south. Meanwhile, the spotted ground squirrel—which is spangled with dots arranged at random, with neither stripes nor rows—is found on dry, sandy grasslands across the south-central plains, from the Nebraska Sand Hills to southwestern Texas.

The rest of the squirrely crew is outfitted in plain brindled coats. Franklin's ground squirrel, a.k.a. the gray gopher, is a largish, plume-tailed rodent that is most at home in thick stands of grass or brush. It ranges along the eastern border of the plains, from Kansas north, and then follows the arc of the Aspen Parklands west into Alberta. The other two species are so similar that until quite recently they were thought to be one and the same. The dapper Wyoming ground squirrel is restricted to three isolated pockets in the western United States, two in the mountains and one smack in the middle of the prairies, in the vicinity of Cheyenne, Wyoming. By contrast, its cousin, the Richardson's ground squirrel, is found across the northern plains, from the Canadian prairies to the central states.

Richardson's ground squirrels—more familiarly known as "flickertails" or "prairie gophers"—look like nothing so much as miniature prairie dogs. The same quivering alertness. The same bolt-upright stance. The same flying flick of the tail as they scoot underground. The similarity between the two species even extends to the intimate details of their living arrangements. Like female prairie dogs, female gophers live in clusters of close kin, seeking safety in the company of their mothers, daughters, nieces, and aunts. The males live on the margins of the colony—or even all alone—in much the same exposed and vulnerable positions as male prairie dogs.

But there is one striking difference between the two species. Unlike ground squirrels, prairie dogs do not hibernate. Although they become dormant for a few days when the weather is extreme, they soon rouse, return to the surface, and resume eating. Ground squirrels, by contrast, go underground in late summer, curl up in a ball, and lapse into a state of utter torpor. Although their tissues occasionally warm themselves, the animals "chill out" for as long as seven months, often sinking to the temperature of the frozen ground. During this whole period, they do not wake or drink or eat; they just lie there conserving energy. This spectacular ability may explain why gophers—and not prairie dogs—are abundant on the winter-bound Canadian plains and have prime importance in the local ecology.

Thirteen-lined

Spotted

Richardson's

Mexican

Franklin's

disperse as juveniles to seek acceptance in another group, but females generally settle down in their natal coterie, close to their mother, sisters, and aunts. Although coterie members defend their space against intruders from neighboring clans, they live companionably among themselves, sharing burrows and other resources—with one significant lapse. During the birthing season, nursing females vigorously defend their nest burrows even against their kin, and for very good reason. The greatest danger to infant prairie dogs is their adult female relatives, who nip into unattended dens and kill the helpless huddles of pink pups. Yet once the pups emerge from their burrows, harmony is restored and adult females take turns nursing youngsters that, only a few days before, they would have attacked with vigor.

The reason for this turnaround remains obscure, but it may have something to do with a seasonal change in the "value" of the pups. In early spring when browse is scarce, another female's offspring may be most attractive as a source of nourishment. But by the time the pups appear above ground, their status changes abruptly from victim to defense. With her own progeny now exposed to predation, each female attempts to draw extra pups around her as protection for her own. If a predator attacks, there is a chance that her offspring will avoid harm by losing themselves in the crowd.

In the last 150 years or so, any threat posed to prairie dogs by their own infanticidal bent has been overshadowed by a greater and more persistent threat. To many cattlemen, a prairie dog colony, with its spreading expanse of close-clipped plants, represents little more than a terrible waste of grass, a view that has traditionally been seconded by range scientists. In 1924, for example, a team of respected researchers stated categorically that the prairie dog "does not possess a single beneficial food habit [beneficial to cattle, that is]; nor is there any argument . . . against its complete eradication on all grazing ranges." With that end in mind, prairie dogs have been poisoned with bait, suffocated with gas, drowned, and shot. Thousands more have been vacuumed out of their dens so that they could be relocated or, more often, offered a "humane" death. The result of this relentless and sometimes gleeful campaign of extermination has been a 98 percent reduction in the population of black-tailed prairie dogs since 1900.

In 2000, the U.S. Fish and Wildlife Department determined that the black-tailed prairie dog merited listing as a "threatened" species under the U.S. Endangered Species Act but determined that "further action to place it on the list is precluded by actions to address higher-priority species." Dissatisfied with

All clear! The prairie dog's "jump-yip" display signals the departure of a threat, such as a predator, from the area. Soon, all the dogs in the vicinity will join in the performance, yipping and jumping at their release from danger.

this ruling, conservation groups petitioned for a new assessment in 2004 and again in 2008, resulting in the species being removed from the list altogether. (The black-tailed prairie dog is listed as a "Species of Special Concern" at the northernmost extension of its range in Canada.) Meanwhile, under state legislation, prairie dogs continue to be targeted as pests, and in some jurisdictions landowners are obligated by law to eradicate prairie dog towns, otherwise known as infestations of varmints.

The war against prairie dogs has been based on the premise that they destroy rangelands. And it is certainly true that a prairie dog colony—with its patches of exposed ground and stumpy, chewed-off plants—bears a striking resemblance to a badly managed pasture. But recent research suggests that while prairie dogs can prevent the plants from recovering by keeping them snipped down, the rodents are rarely the primary cause of overgrazing. Because they cannot move into tall stands of dense vegetation, they can only follow where heavy grazing by cattle (or formerly bison) has opened up a path. Ranchers who overtax their rangelands have effectively laid out a welcome mat and invited the prairie dogs to "infest" their land.

The claim that prairie dogs strip the prairie clean and leave nothing for larger herbivores to eat is also incorrect. In fact, far from avoiding prairie dog towns, bison are attracted to them and, given the chance, will spend a disproportionate amount of time grazing on the rodents' well-maintained lawns of tender young blue grama and buffalo grass. Although prairie dogs do eat plants that bison, or cattle, might otherwise consume, they compensate, at least in part, by maintaining patches of highly nutritious, highly digestible fodder. Presumably this fact explains why cattle that share the range with prairie dogs may gain weight almost as fast as those that have the pasture to themselves. Although the arithmetic varies from case to case, the monetary reward for removing prairie dogs is often offset by the expense of "controlling" them.

The ecological benefits of sustaining prairie dogs, by contrast, are incalculable. Prairie dogs are the beavers of the grasslands, in the sense that they reengineer the environment and create living space for an astonishing variety of other creatures. Through their constant gnawing, they not only maintain fresh fodder for cattle and bison but also increase the diversity of the vegetation by opening up space for forbs such as scarlet mallow (or globemallow, as it is also called) and prickly pear cactus. These plants, in turn, provide food and shelter for other herbivores, including insects, birds, mice, rabbits, and hares. Some of these animals—cottontails, for example—are also attracted by the shelter of the prairie dogs' mounds, where they can tuck themselves down out of the sun for hours at a time. Burrowing owls just move right on in, laying claim to abandoned burrows (usually on the outskirts of town) and building their nests underground. In the hazy heat of midsummer, the little owls can sometimes be seen standing, swivel-necked, at their burrow mouths or flapping languidly over the colony in search of grasshoppers.

Black-footed ferret

More than 160 species of vertebrates have been observed on prairie dog colonies. Although some are drawn by the opportunities that prairie dogs create, others are attracted by the prairie dogs themselves. The plump little rodents figure in the diets of a host of predators, including two species of rattlesnakes and a dozen different kinds of hawks, plus weasels, skunks, badgers, bobcats, coyotes, and foxes. Prairie dogs also are, or at least once were, the sustaining resource of what is now one of the rarest carnivores on Earth. The black-footed ferret—a sinuous, mink-sized weasel with a black-bandit mask—is built to slink through prairie dog burrows and capture the rodents when they least expect it. Not surprisingly, as prairie dog populations dwindled, the ferrets also became increasingly rare. To make matters worse, the ferrets are also subject to

Under the watchful glare of a parent, right, two young burrowing owl chicks venture out of their den.

Arthur Savage photo

canine distemper and to plague, an introduced disease that is endemic in prairie dog colonies. Under this malign confluence of influences, the species has come within a heartbeat of extinction. By 1987, only eighteen individuals were known to exist, in a single prairie dog town in Wyoming. Now, all these years later, the descendants of this remnant population are being bred in captivity and reintroduced to selected prairie dog towns across the plains, from Saskatchewan to Mexico. Although their future remains precarious—the threat of disease is always there—the population has grown from essentially zero to around a thousand.

Pronghorns: Outwit, Outplay, Outlast

Since the moment when Brewster Higley penned his verses on the banks of Beaver Creek, "discouraging words" about the future of the prairie's unique plants and animals have been heard all too frequently. In the last quarter century, for example, burrowing owls have essentially disappeared from the Canadian prairies, despite concerted efforts to sustain them, and declines have also been detected in many parts of the American plains. Similarly, the populations of mountain plovers, pale killdeerlike shorebirds that nest right out on the bald-headed prairie, often on prairie dog towns, have plunged by more than 60 percent in the last forty years. Much-loved birds, like prairie chickens, have been declining steadily, and signature species, like burrowing owls, have also been losing ground. "Less" and "fewer" have become our watchwords.

Happily, some species have bucked these dismal trends; the pronghorn is chief among them. Once as fantastically abundant as bison, with an estimated precontact population of 35 million head, the species was decimated by over-hunting during the settlement era. By 1900, the herds had been reduced to a tattered remnant of some 15,000. But a determined conservation effort turned the situation around, and the herds have rebounded several-hundred-fold in the last century. The count currently hovers around 1 million animals.

The pronghorn is a survivor. Although dozens of related species once roamed the Great Plains, only one, *Antilocapra americana*, made it through the mass extinctions at the end of the Ice Age. In so doing, the pronghorn permanently outdistanced its most fearsome predators, including North American lions and dire wolves; yet even today, it flies across the prairie as if they were still in hot pursuit. The fastest animal in North America, and possibly in the world, it can hit speeds of 60 miles (100 kilometers) an hour and maintain a blistering pace for a couple of miles. Running open mouthed, it gulps air down its extralarge windpipe and into its extralarge lungs, as its powerful heart keeps its muscles fueled with oxygen. This is a species with the prairie in its blood, the animal world's answer to wide open spaces.

Even the normally stolid Captain Clark—of Lewis and Clark fame—was moved to poetry by the pronghorns' grace and speed, which put him in mind of "the rapid flight of birds." Other observers were reminded of gazelles and jumped to the conclusion that pronghorns were a kind of antelope. (In fact, antelope are bovids, like bison and cattle; pronghorns are the only members of their own unique family and genus.)

Naturalists have found the pronghorn's social life even more deceiving. Although pronghorns of all ages and both sexes congregate during the winter, the herds break up in the spring, as the animals spread out across the prairie. Mature males—the big guys with the forked black horns—often lay claim to one or two sections of rangeland, which they defend against other bucks. Meanwhile, the does—the ones with the straight, short horns or occasionally none

at all—gather in small herds that move unhindered from one male territory to another. Or so it goes until late summer when, with the approaching rut, the mature males begin to aggressively herd the does, as they stake sexual claim to their "harems." At least, this is what early observers thought was going on.

What actually happens turns out to be far more mischievous. Although the males do their best to keep tabs on the does, they are not always successful. If a female can't break away from the buck with a feint and a bound, she may outwit him by urinating in front of him. Distracted by her deposit of intoxicating scents, he indulges in flehmen, or intense sexual sniffng, while she makes a run for it. In the weeks just before estrus, a female may visit several breeding males before consenting to mate, apparently basing her decision on the bucks' relative ability to protect the herd from harassment and their success in male-to-male combat. Sometimes a female sneaks away from the herd just long enough to attract a consort of nonterritorial, or bachelor, males and returns to parade them past the harem buck. Standing safely out of harm's way, she then watches the snorting, pawing, rushing, and chasing that ensues, before presenting herself to mate with the victor. Far from being complete masters of the situation, male pronghorns are in much the same position as grouse on a lek as they strut their stuff before a discerning audience.

With the hairs on his rump flared in alarm, a pronghorn flashes a warning to the rest of the herd as he darts away from danger.

If the hidden quirks of pronghorn society are only beginning to be understood, one aspect of their biology has been clear from the outset. Although pronghorns are native to grasslands, they do not eat very much grass. As a result, they are perfect companions to grass-eating species such as bison and cows and do not interfere with the commercial production of cattle. In lieu of grass, pronghorns prefer a diet of shrubs and forbs, using their small, dainty muzzles to nip off the choicest morsels. This ability

Prairie sage

Silver sagebrush

facing page: Once an animal unique to the West, the coyote has expanded its range to become the most widely distributed wild dog in North America. In recent years, it has moved into city parks and green belts, establishing itself as part of the urban scene.

to forage selectively, one mouthful at a time, is important because many of the pronghorns' favorite foods are laced with poison. In the long evolutionary struggle between herb and herbivore, plants have fought back with every weapon at their disposal. Just as grasses contain grit that wears down herbivores' teeth, so many grassland forbs contain toxins that can cause indigestion, liver damage, miscarriage, and other bodily harm. (Locoweeds, lupines, and larkspurs are just three of a long list of prairie plants that regularly kill cattle.) But through evolution, many prairie herbivores have responded to this threat by developing the ability to neutralize certain of these toxic agents. For example, sage and sagebrush contain turpentinelike compounds that if ingested and left unchecked, interfere with the microorganisms in an ungulate's complex stomach. Cattle avoid sages, and even mule deer can't eat too much, but pronghorns have completely mastered the challenge. Sage and sagebrush are among their staple foods and feature in their diet throughout the year.

Wile E. Coyote and Co.

The first rule of ecology, on the rangelands as everywhere else, is that everything and everyone eventually gets eaten. But because organisms expend so much energy in daily living, only a fraction of what they ingest is converted into living tissue, or biomass. The proportion that is available for transfer to the next trophic level is usually no more than 10 percent. As a result, it takes a vast span of grass to support a herd of herbivores, which in turn can sustain only a handful of predators. Yet despite their relative scarcity, predators are present on the rangelands in every conceivable size and shape, from eight legged to six legged to no legged. Think of spiders, tiger beetles, ladybugs, assassin bugs, robberflies, and wasps; think snakes, lizards, turtles, frogs, toads, owls, and hawks; think shrews, weasels, skunks, and wild cats. Time was when the catalog would also have included wolverines, cougars, grizzlies, American black bears, and wolves, but those species were pushed off the plains during the settlement era. Wolves in particular were targeted for extermination with strychnine, traps, and guns, as an insult to commerce and a threat to civilization. (Today, although young wolves sometimes migrate across the prairies from the mountains and the north woods, they are inevitably shot or hit by a car before they have a chance to settle in.)

In the Bad Guy/Good Guy paradigm of the nineteenth century, the eradication of wolves should have been pure gain. Instead, it merely opened the way

for a new breed of "outlaw" predator to move into the vacated space. As long as wolves were the dominant canids on the prairie, their slightly smaller cousins, the coyotes, didn't stand much of a chance. Any coyote that entered a wolf territory was likely to meet a bloody end, a victim of the cold logic of interspecific competition. As a result, coyotes were generally restricted to the gaps, or buffer zones, between packs of wolves and, though widely distributed across the plains, were nowhere plentiful. At least that is the impression one gets from the accounts of early explorers, who seldom mention seeing coyotes or, to use their

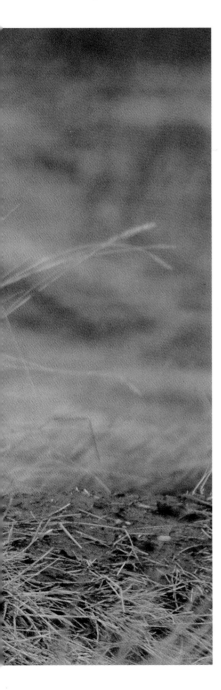

phrase, "bush wolves." By contrast, both red foxes and the diminutive (pussy-cat-sized) swift foxes appear to have been widespread and common in some localities, perhaps because they were too small to compete with wolves and thus were allowed to go their jaunty way without interference.

After wolves were removed from the scene, the coyotes muscled in, and the balance of power among canids began to shift. Coyotes are versatile predators. Depending on the circumstances, they can band together in packs to take deer and compete with wolves, or they can hunt alone to catch rodents, rabbits, and other small prey much as foxes do. So just as wolves had once impinged on coyotes, coyotes now aggressively took steps to limit the populations of their smaller competitors. Red foxes, in particular, took a major hit, and by 1900 had essentially vanished from large areas of the grasslands. As the agricultural landscape changed, from the 1930s on, red foxes pushed back in from the east and reoccupied the region.

A super-abundance of coyotes may also have been a precipitating factor in the decline of the swift fox. The smallest of the wild dogs, these guileless little creatures were once common across the short- and mixed-grass prairies, from Canada to Texas and from the Rockies to the Dakotas. By the 1940s, however, the species had been extirpated from the northern plains, probably as a result of coyote predation, together with loss of prey (notably rodents), loss of habitat to the plow, and loss of life to poison baits intended for other predators. Thanks to decades of effort by conservationists, small populations of swift foxes have recently been reestablished on the Blackfeet Reservation in Montana, in southern Alberta and Saskatchewan, and

A red fox kit, only a few weeks old, explores the world around the den in which it and its littermates were born. (As a rule, both coyotes and red foxes bear litters of six to eight, while swift foxes average only three or four.)

Coyotes and American badgers are middle-sized carnivores that share a taste for ground squirrels, prairie dogs, and other middle-sized prey. One might expect, then, that prey taken by one species would be a loss to the other. The fact that badger populations increase when coyotes are removed suggests that this kind of competition actually does occur and that badgers are typically the losers.

Under certain circumstances, however, the rules of engagement appear to shift from winner versus loser to win-win. If the two species find themselves working the same rodent colonies, and if such prey is scarce, the hunters may form an unlikely alliance. As the badger pursues its prey by burrowing underground, a coyote stands guard up top, ready to pounce on any rodent that makes a run for it. Meanwhile, down below, panic ensues, and the hardworking badger takes advantage of the confusion to make its kill.

Although hunting partners never share food with each other, they sometimes take time out to play—even to wrestle like pups—or simply to touch noses or lie down with each other. One observer described an encounter in which a coyote laid its head on the badger's neck, licked its face, jumped in the air, and made "other expressions of unmistakable joy." No wonder the two predators are often described in Native stories as neighbors and friends and are even said to address each other as cousins.

in South Dakota. In Canada alone, the population has risen from zero to around 650 since 1983, when the first captive-bred animals were released. These days, the greatest impediment to the survival of swift foxes appears not to be direct predation by coyotes (though, of course, that still occurs) so much as competition with their canny red cousin. When red foxes are present, swift foxes make themselves scarce, though exactly why this happens, no one can say for sure.

Swift fox

One thing we can be sure of, however, is that Wile E. Coyote still reigns supreme as top dog across the prairies. After a century or more on agriculture's Most Wanted list as livestock predators, coyotes are still out there everywhere you look, howling, breeding, birthing, going about their business. By any sensible measure, attempts to limit their numbers have been a complete flop, much like trying to empty the bathtub without turning the water off. To have any lasting effect, a coyote-control program must target at least 75 percent of the breeding animals in a given neighborhood, a goal that is seldom achievable in practice. Yet without such drastic reductions, "control" becomes merely a means of thinning the population and ensuring that the survivors have adequate food to put them in prime condition to reproduce. When food is abundant, mature females produce large litters, and even subadults under the age of two are likely to reproduce and make a contribution to the baby boom.

But if coyote "control" can't be relied upon to achieve its intended results, it inevitably triggers a cascade of subtle, unseen, and unanticipated changes. For example, a reduction in the numbers of coyotes (however temporarily) provides an opening not only for red foxes but also for skunks, bobcats, and other middle-sized predators, many of which prey heavily on duck nests. Fewer coyotes thus translates indirectly into fewer ducks and a chain of unknown effects on wetlands.

Meanwhile, on the grassy uplands, the removal of coyotes can also cause unexpected shifts in the abundance of rodents. For instance, one study in Texas found that when coyotes were present in good numbers, the prairie was home to flourishing populations of a dozen species of microherbivores, including deer mice, pocket mice, harvest mice, and Ord's kangaroo rat. But after half of the coyotes had been shot, the mouse populations crashed and the only species that remained was the kangaroo rat. (Presumably, predation by coyotes normally kept kangaroo rats in check and allowed the other, less competitive species to claim living space.) Since kangaroo rats are aggressive foragers that can push grasslands over the brink and encourage the invasion of desert shrubs, their dominance may have made the rangelands less hospitable for cattle. Meanwhile, with the coyotes held in abeyance, black-tailed jackrabbits multiplied to such an extent that they consumed any cattle fodder that was left. Thus, far from being dispensable animals (as one range manager described them in 1973), coyotes are a keystone species in rangeland ecology and the unlikely guardians of the cattle industry.

Long honored as the Trickster, Coyote still walks the prairie with us, calling us to look beyond the obvious.

six

WATER OF LIFE

Here is a land where life is written in water.

THOMAS HORNSBY FERRIL, COLORADO, 1940

IT WAS LITTLE more than a puddle at the side of the road, the kind that appears in spring and is gone by midsummer. Two days earlier, it had lain frozen, burdened by a freak spring storm. But in the sudden warmth of late April, the tide of the seasons had turned and the hollow was filled with sunlight and the bright creaking of frogs.

The far bank bristled with willows and wild roses and, poking up out of their midst, a single, stunted aspen that carried two large masses of twigs. One was flat topped, like a crow's nest, and did not appear to be in use, but the other was massive and messy and comically overbuilt for the meager tree that supported it. It had to be a hawk's nest. And, yes, there was a bird on this one, a large, dark form with a knob of a head that popped up above the rim. Do you suppose that it's a Swainson's hawk—the kind with the rich chestnut-brown bib—or its rarer cousin, the pale-fronted ferruginous hawk? Out come the field guides and binoculars. But wait: the shape that is sprawled on the hawk's nest doesn't look like a raptor at all. It's some kind of gray-backed bird with a sinuous black neck, a bar of white on its cheek and a stubby, ducklike bill that extends languidly over the rim of the nest. Could it be—what else could it be? A Canada goose!

I can't tell you if that goose had merely settled on the nest to sunbathe or if it had inexplicably taken a notion to lay and incubate its clutch at the top of a scrubby aspen. Strange as it

There was a world before this world, the Lakota sto-rytellers say, but the Creating Power was not pleased with it. He said to himself, "I will make a new world." So he took his sacred pipe and his pipe bag, and he sang four songs. When he sang the first song, it started to rain. When he sang the second and third songs, it poured, and the rain-swollen rivers over-flowed their beds. When he sang the fourth song and stamped on the earth, it split open in many places like a shattered gourd, and water flowed from the cracks until it covered everything.

By the time the rain stopped, all the people and animals had drowned. The Creating Power thought: "It's time to unwrap the pipe and open the pipe bag." The wrapping and the bag contained all manner of animals and birds, from which he selected four ani-mals known for their ability to stay under water for a long time.

First he sang a song and took the loon out of the bag. He commanded the loon to dive and bring up a lump of mud. The loon did dive, but it brought up nothing because the water was too deep. The otter and the beaver tried in their turn, but they too brought up nothing.

At last the Creating Power sang the fourth song and took the turtle out of the bag. "You must bring the mud," the Creating Power said. The turtle dived into the water and stayed below so long that the other three animals were certain it was dead. But when it finally surfaced, its feet and claws—even the space between its upper and lower shells—were filled with mud.

Scooping mud from the turtle's feet and sides, the Creating Power began to sing. He sang as he shaped the mud in his hands and spread it on the water to make a spot of dry land. Eventually, he made so much land that the water was replaced by earth. "Water without earth is not good," thought the Creating Power, "but land without water is not good either." Filled with pity, he wept for the earth, and his tears became oceans, streams, and lakes. "That's better," he thought . . .

He named this new world Turtle Island because it was made from the mud that the turtle had brought from under the water. (This is an abbreviated version of the story as it was told in the early 1900s.)

may seem, this turns out to be a credible possibility. Although Canada geese usually nest on the ground near water or, at most, on slightly elevated surfaces such as grassy hummocks and beaver lodges, they actually take over abandoned hawk nests as much as 6 yards, or meters, above ground. When the goslings are ready to leave, they step off the nest platform and flutter their tiny wings toward what will hopefully be a soft landing. Be that as it may, there is one thing I do know for sure. Even the most unprepossessing prairie puddle is a site of wonderment, a place where the unexpected happens daily and where nothing can safely be taken for granted.

It isn't just the life of watery places that is so interesting. Water itself is amazing. For instance, the droplets that collect in a roadside ditch, or that rain down on our heads, may have been in existence for billions of years, perhaps since the earliest beginnings of the universe. Originally derived from the cloud of roiling, boiling gases that gave rise to the sun and stars, water is a fusion of hydrogen and oxygen into a compound, H_2O, that still participates in the restless energy of creation. It is constantly in a process of transformation. That raindrop on your nose was, until recently, a cloud. And before that, those same molecules of water may have gone through countless passages from snowfall to spring run-off, groundwater to marsh, marsh to river, river to ocean, ocean to wind, wind to cloud, in an endless, cyclic journey through the physical world.

And that is still not the whole picture. For not only is water capable of shifting from solid to liquid to gas—making it the only substance to exist in all three states under normal conditions on Earth—it also routinely makes the unfathomable leap from nonliving to living. The miracle of life derives from a thin, protoplasmic soup that, depending on the function of the cell, consists of between 50 and 90 percent water. (Our own tissues average out at about 50 percent.) So that same raindrop, which by now is dribbling off your chin, has likely also made a journey through the living world. From rain to soil, root to leaf, leaf to goose, goose droppings to soil, soil back to air, it has flowed through the food web, moving freely from organism to organism. Has it experienced life as an amoeba? An earthworm? A brontosaurus in a swamp? In the shape-shifting world of water, stranger things have happened.

Water is the medium of life, so it is not surprising that aquatic habitats—including oceans, rivers, lakes, ponds, and marshes—are among the richest and most productive environments on Earth. But it may be surprising to discover that this generalization holds true for the landlocked Great Plains, a region that we typically associate with drought, dust storms, and water shortage. On a planet where two-thirds of the surface is covered by water, the wide expanse

Even in midwinter, a film of algae brings color and life to this pond in the Black Kettle National Grassland, near Cheyenne, Oklahoma.

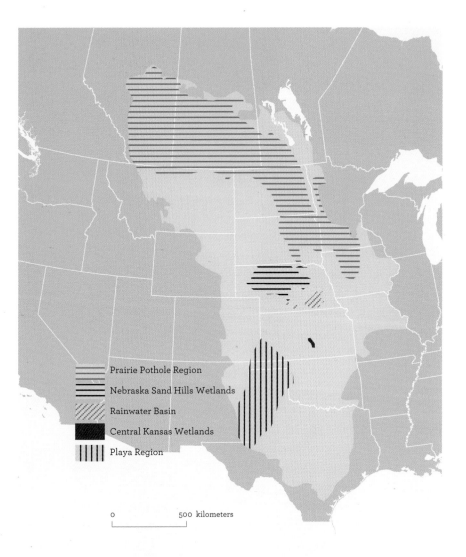

Prairie Pothole Region
Nebraska Sand Hills Wetlands
Rainwater Basin
Central Kansas Wetlands
Playa Region

0   500 kilometers

WETLAND REGIONS OF THE GREAT PLAINS

of the prairies is a sea of dry land, broken here and there by small islands of aquatic habitat. Broadly speaking, the area covered by water is greatest in the eastern, tall-grass country and diminishes toward the west, following the logic of declining precipitation. Thus, in parts of central Manitoba, the inrush of moisture-bearing air from both west and south produces lush meadows and extensive, reedy marshes that occupy as much as 25 percent of the landscape. But on the prairies of Alberta, where the mountains block out the rain, wet

meadows and marshes cover a mere 5 percent of the range. Similarly, if you cross the border into the States and head for the southern plains, the area of surface water declines incrementally. In Nebraska and Texas, for example, where the summer sun has the power to suck whole rivers out of their beds, the coverage of water declines to a fraction of a percent. Yet, though limited in physical area, the aquatic habitats of the prairies are biologically immense, capable under ideal conditions of producing more life in less space than almost any other environment on the planet.

Putting Marshes on the Map

For the purposes of discussion, it is useful to divide the aquatic habitats of the Great Plains into two categories: wetlands, or marshes, and rivers and lakes. Scientifically speaking, the term "wetland" refers to any shallow depression—less than 6 feet (2 meters) deep—that is capable of holding water for long enough to permit the growth of sedges, cattails, and other nonwoody, semiaquatic

The rising trill of the red-winged blackbird is as bright, in its own way, as the epaulettes on the wings of breeding males. Redwings are found in wetlands across the Great Plains, where they build and defend nests in dense stands of cattails.

Arthur Savage photo

vegetation. In other words, it is basically a puddle filled with water-loving plants. Yet this deceptively simple definition leaves room for almost infinite variability in water depth, temperature, permanence, chemistry, soil type, fertility, life history, and just plain chance, so every prairie wetland is, to some extent, a unique habitat that supports a unique community of organisms.

Millions of these mostly small, self-contained marshes are dotted across the Great Plains, providing resource-rich refuges for a remarkable diversity and number of insects, reptiles, amphibians, mammals, and—most spectacularly—birds. For example, the Llano Estacado, or High Plains region, south of Amarillo, Texas, is so emphatically dry and flat that it cannot sustain the flow of a permanent river. Once described as the "great Zahara of North America," the region is now essentially a wall-to-wall expanse of cotton fields under irrigation. Yet this dreary prospect is redeemed by the presence of around 20,000 shallow, saucer-shaped depressions known as playas, or playa lakes, that serve as havens for wildlife. (An additional roughly 6,000 playas occur farther north in the Texas Panhandle, as well as in adjacent parts of New Mexico, Colorado, Kansas, and Oklahoma.) Scoured out by some still-mysterious interaction between wind, water, and underlying geology, these clay-lined basins have been in existence for at least ten thousand years and, judging from remains found buried in their beds, once served as watering holes for both Columbian mammoths and the Folsom-era humans who hunted them. Somewhat more recently, in 1541, when Francisco Vásquez de Coronado marched across these plains, he was as impressed by the playas—"round like plates, a stone's throw wide"—as he was by the bison that he encountered near them.

Throughout their long history, playas have presumably functioned in the same way as they do today: as catch basins for runoff and rain. As a rule, they fill up during spring cloudbursts, reach peak levels in June, and thereafter (through evaporation and slow seepage) gradually disappear. The deeper the depression and the more water it holds, the longer the playa persists through the season. Since most playas are flat bottomed and less than a yard in depth, they are generally dried out by December, at the latest. Only the deepest basins—including the increasingly large number that have been modified to retain water for irrigation—hold moisture year round and during periods of drought. Yet despite their ephemeral nature, playas sustain an abundance of life, including not only semiaquatic plants such as barnyard grass, curly dock, and pink smartweed but also a mix-and-match selection (differing from pond to pond)

facing page: Mayflies, genus Ephemera, spend most of their lives in the water, first as eggs, then as herbivorous larvae that scuttle and scurry along the bottoms of streams. The airborne adult form survives for no more than a day or two, just long enough to reproduce.

of up to six species of waterboatmen; five types of dragonflies; fifteen kinds of small marsh flies, or midges; nineteen kinds of freshwater shrimp; and dozens of other squirming, swimming, and flying invertebrates. This living soup, in turn, sustains populations of larger organisms, including amphibians (up to fourteen species of salamanders, frogs, and toads), reptiles (especially yellow mud turtles), and mammals (deer mice, cottontails, badgers, striped skunks, coyotes, and raccoons, among others).

But playas, above all, provide a haven for birds. During the breeding season, these unassuming wetlands serve as oases of nesting habitat both for marsh-adapted birds, such as ducks and red-winged blackbirds, and for a wide variety of typically terrestrial species, including short-eared owls, western meadow-larks, and ring-necked pheasants. The region also provides migratory stopover sites and wintering habitat for enormous numbers of waterfowl and shore-birds that breed elsewhere on the continent. Roughly 2 million ducks and geese crowd onto the playas each fall, including a gabbling contingent of mallards, northern pintails, American wigeons, and green-winged teal that nest in west-ern Canada and the Dakotas. What's more, playas provide off-season accommo-dation to flight upon flight of gurgling sandhill cranes—more than 300,000 in all, or 90 percent of the midcontinent population—which land on the mudflats and in the dwindling ponds to feed on the thick slurry of crustaceans. And this is still to say nothing of the stilt-walking American avocets, with their upturned bills, or the Wilson's phalaropes, which spin round and round like tops to stir up food.

Little understood and vastly underappreciated, the playas of the southern High Plains are a vital ecological resource that helps to sustain waterfowl and shorebird populations over much of the continent. This function is shared by a number of other equally important wetlands in other parts of the Great Plains, many of them, like the playas, out in the middle of the prairies. The Central Kan-sas Wetlands, for example, include a number of freshwater and saline marshes—notably Cheyenne Bottoms and the Quivira National Wildlife Refuge, both near Great Bend—that lead like stepping-stones through what is otherwise a vast plantation of corn, sorghum and soy. Located on a major midcontinental migration route, these wetlands annually play host not only to waterfowl by the millions but also to half of all the shorebirds in North America. Although significant populations of shorebirds breed in Kansas and elsewhere on the Great Plains, a much larger number beat northward to the Arctic coast, only

Short-eared owl

facing page: When feeding, the American avocet thrusts its bill underwater and swings it from side to side along the bottom, stirring up a flurry of tasty aquatic insects.
Arthur Savage photo

Wilson's phalarope

> ### BIRDS ON THE RUN

Shorebirds, those long-beaked, long-legged stilt-walkers that feed on mud flats and shallow marshes, can be a tricky group for a bird watcher to bring into focus. They are like shimmering apparitions, caught in a dazzle of light and water. Almost frantically active, they peck, they probe, they wheel up and circle over our heads, sometimes uttering strange, quavering cries in their own strange languages. And then, faster than you can say, "black-bellied plover," most of them have gone, either headed north to breeding grounds on the tundra or, a few short weeks afterward, flying south to far-distant climes for the winter.

Of the thirty-seven species of shorebirds that have been sighted on the Great Plains, the great majority—twenty-three species, mostly sandpipers—were migrants noted in passing. Several of them, like the tawny little Baird's sandpiper—no larger than a bluebird—make an annual round-trip journey of up to 20,000 miles (32,000 kilometers) from the High Arctic islands to the outer limits of Tierra del Fuego. Although Baird's and several other species cut across the prairies on both legs of the trip, others—including the even tinier but otherwise confusingly similar semipalmated sandpiper—follow an elliptical route that crosses the interior of the continent on the northward journey and loops east to the Atlantic Coast for the return. In yet another variation, another close cousin, the western sandpiper, goes through this same pattern only in reverse, traveling north via the West Coast in spring and heading south across the Great Plains in the autumn. The result is a constantly changing playlist of shorebirds on prairie marshes and sloughs, as the sandpipers of the continent pass by us in revue.

The remaining fourteen species of shorebirds, by contrast, are not driven to such extremes. Mainly short-to-medium-distance migrants, they tend to winter on the coasts of North or Central America and to nest in the temperate zone, including the wetlands of the Great Plains. This group includes such familiar birds as the killdeer, with its keening cry and broken-wing display; the American avocet and long-billed curlew, with their adroitly curved beaks (the former for scything through the water, the latter for probing); and the spotted sandpiper, the only "peep" that, as its name suggests, can easily be identified at first glance, at least during the breeding season. Spotted sandpipers are also noteworthy for their practice of polyandry, a mating system in which the male, as Mr. Mom, incubates and rears the nestlings, while the female freely pursues other matings and produces several broods.

to breed, turn around, and head back south, often to wintering grounds on the other side of the equator. Hence, the critical importance of refueling stops along the route.

Just north of the Central Kansas Wetlands lies the Rainwater Basin of south-central Nebraska, a rolling landscape dotted with several thousand playa-alike catchments. And then it's on to the Platte River, with its sandbars and wet meadows, and, a little farther along, the clear, sparkling—utterly incongruous—ponds that spangle the Sand Hills of northwestern Nebraska. Filled to the brim by the Ogallala Aquifer (the vast reservoir of fossil water that underlies the south-central plains) and set against the silence of the grassy dunes, these wetlands are bursting with life and sound: the high-pitched complaints of long-billed curlews, the peeping of sandpipers, the metallic rattle of marsh wrens at nests in the cattails.

Long-billed curlew

The final entry on the list of major wetland resources on the Great Plains is the Prairie Pothole Region, an area of some 270,000 square miles (700,000 square kilometers, about the size of Texas) that angles northwest from north-central Iowa through western Minnesota and the Dakotas and across the Canadian prairies from Manitoba to western Alberta. This entire swath of country is pocked with millions of mostly small, knee- to waist-deep depressions—called potholes, or sloughs—that are a legacy of the most recent glaciation. As the glaciers decayed, chunks of ice became buried at random under layers of gravel and silt, and when these blocks eventually melted, the ground collapsed, turning ice mounds into small meltwater basins. In keeping with this Ice Age heritage, prairie potholes still depend not so much on rainfall as on snowmelt, trickling over frozen ground, to replenish their supply of water.

Prairie potholes are proof of the adage that every little drop counts. Insignificant if considered in isolation, they gain significance when seen as elements of a sweeping, if fragmented, aquatic complex. The combined area occupied by potholes in the northern states, before settlement began, is estimated to have been 20 million acres (8 million hectares), giving them the same collective presence on the landscape as Lake Superior. And since 80 percent of the pothole region lies in Canada, this comparison barely hints at the wetlands' original extent, much less at their ecological significance. Vitally important as a source of protein for both resident and migrant birds, these weedy, reedy marshes are perhaps most famous as the "duck factory" of North America. Although they only amount to about 10 percent of the waterfowl-breeding habitat in North

America, they typically produce between 50 and 90 percent of all the ducklings that fledge in any given year. Clearly, prairie potholes are spectacularly productive. The obvious question is, how do they do it?

How Prairie Potholes Work

At first glance, a slough does not appear to have the makings of a biological powerhouse. What is it, in the final analysis, but a pool of stagnant water? Sunk into poorly sorted glacial rubble that inhibits organized flow, a typical pothole has only weak and fitful connections with the surrounding groundwater. Depending on the circumstances, any given pothole may receive water through seepage, lose it through underground leaks, or experience both input and outflow through different parts of the basin simultaneously. A slough that receives an influx of groundwater is also typically hit with a burden of dissolved salts picked up from the underlying geological formations. If this briny solution is permitted to escape from the basin either through overbank flooding or underground leaks, the salts flow out with the water and tend not to accumulate. But if the basin has no outlet, the water gradually evaporates, the concentration of salts builds up, and the wetland becomes saline (with chlorides) or alkaline (with sulfates). Some prairie potholes are ten times as salty as the sea, surely pushing the limits of habitability.

Sloughs also present a variety of other potentially life-threatening challenges. In the swampy warmth of midsummer, as submersed plants die back and rot, the water becomes starved for oxygen—if there's still water left to starve. Like playas, prairie potholes are notoriously unstable, typically filling in spring and dwindling over a period of weeks or months, with little chance of renewal before the next spring thaw. How long water persists varies from pothole to pothole and from year to year, in response to such variables as the depth of the basin, flow to and from the groundwater, and the uncertainties of the weather. Depending on its duration, a slough may be classified as ephemeral, that is, a thin layer, or sheet, of water that lasts for a matter of days; temporary, a puddle that persists into early summer; seasonal, a pond that dries out annually but that often remains wet till fall; or permanent, a wetland that withstands all but the most severe droughts. (This last category includes such birding hotspots as Delta and Oak Hammond marshes in Manitoba, Last Mountain Lake and the Quill Lakes in Saskatchewan, and Beaverhill Lake in Alberta, all of which have been honored as Wetlands of International Importance under

the UN's Ramsar Convention on Wetlands.) And these distinctions are all overridden by the insult of the northern winter, which closes in on wetlands regardless of their status and deprives them of light, warmth, and other necessities.

Yet the wonder is that, in one way or another, life goes on. For example, lack of dissolved oxygen is a crisis for many species that breathe through their skin (like flatworms) or with gills (like dragonfly larvae, or nymphs) and that accordingly are restricted to only the most permanent and well-aerated sloughs. But certain other species are adapted to cope with anaerobic or near-anaerobic conditions. Some daphnia, or water fleas, for instance, respond to oxygen-starved water by pumping up their production of oxygen-grabbing hemoglobin, an adaptation that not only turns them a surprising shade of pink but also enables them to take full advantage of every available molecule of dissolved oxygen. Mosquito larvae, or wrigglers, by contrast, live in the water but obtain their oxygen by means of breathing tubes that poke into the air, on the same principle as

One of the largest shorebirds in North America, the marbled godwit feeds by probing the mud with its long, slightly curved bill, searching for mollusks, crustaceans, and worms. It nests on grassy prairie near the edges of sloughs and alkaline lakes in the southern Prairie provinces and the northern plains states.

snorkelers. Similarly, a variety of aquatic bugs, such as waterboatmen and back-swimmers, rise to the surface and pick up loads of air, which they carry down with them like scuba divers. Often you can see them scooting around beneath the surface of the slough, all shiny and silvered with bubbles.

So far so good. But what happens when the very medium of pond life—the water—disappears? This is a routine crisis for prairie potholes (and other transient wetlands on the Great Plains) and one that, to a remarkable extent, evolution has turned to advantage. Through thousands of reiterations, the so-called wet-dry cycle has given rise to an ecosystem that not only is adapted to cope with severe hardship but depends on it. Here's how it works. If a basin remains full of water for even two or three years in a row, the community of life within it begins to deteriorate, becoming both dramatically less diverse and less productive. The first sign of this degeneration is a gradual die-back of cattails, bulrushes, and other emergent growth—those plants that are rooted in the muddy bottom, push up through the pond, and emerge crisp and spiky above the surface. With continuous inundation, this once-vigorous foliage either disappears from the basin over a period of months or else recedes toward the margins.

Although the causes of this die-off are often mysterious (aging rootstocks or lack of oxygen in the mud?), the process is sometimes accelerated by an assault force of muskrats. Where conditions are suitable—namely, in well-vegetated potholes that retain free water throughout the year—these plump little herbivores often increase to such extraordinary numbers that they "eat out" large areas of marsh vegetation. (They are basically aquatic field mice and just as fecund.) Their habitat destroyed, the rodents also crash, leaving the wetland doubly diminished.

One loss leads to the next. Fewer emergent plants means less habitat for all the little creatures that live on and around their leaves and stems, including algae, tiny crustaceans, snails, and insect larvae. In addition, the elimination of emergent vegetation allows the wind to stir through the pond, thereby muddying the water and reducing the amount of light available to plants that grow beneath the surface, including the slimy, undulating stems of sago pondweed and coontail. As this submersed vegetation begins to die off in its turn, the marsh completes its degeneration phase and reaches its lowest ebb—the lake phase—a stage that is characterized by large areas of open water and relatively low production of plants, microorganisms, insects, and everything that eats them. And what doesn't?

Dragonfly and
emerging nymph

If you've ever seen swarms of spidery-looking insects zooming across a ditch or pond, then you have made the acquaintance of water striders. Members of the order Hemiptera, or "true bugs," they are one of a small number of organisms specialized to live not in the water but on the water's surface. Like skaters scooting over a frozen pond, they dart around on quiet streams and ponds, so light and quick that they scarcely make a dent in the surface tension. Buoyed up by a cushion of air trapped on the underside of their bodies in a layer of water-repellent hairs, they propel themselves forward with a jerky, oar-like motion of their spindly middle legs, while their long hind legs angle out to the rear for stability and steerage.

Although some species of water striders can fly and all can leap a short distance into the air, their lives are essentially focused on the two-dimensional surface. For instance, the diet of a water strider typically consists of terrestrial and aquatic insects and other small animals that either strike the surface or are unfortunate enough to be trapped in the surface tension. Like a spider in a web, the strider monitors the pond for ripples that indicate a potential meal is within reach. With a few deft strokes, it then spurts to the scene, grasps its prey with its clawed front feet, and stabs its sucking mouthparts into the body.

Water striders also use the water's surface as a kind of drum for social communication. By jarring the water with their legs, they beat out rhythms at different frequencies, thereby conveying different meanings. For instance, a male water strider may send out a quick rat-a-tat to keep other adult males away from its mate but shift to a slow, steady pulse when it is courting. This "good vibration" seems to motivate the female to lay her fertilized eggs, which she attaches in neat parallel rows to floating leaves or stems just beneath the surface of the water.

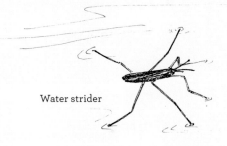

Water strider

Daphnia

Waterboatman

Backswimmer

The slough persists in this depressed condition until it is revived by drought. To regenerate, it first has to dry out or, as the biologists put it, experience drawdown. Although the plants that are at home in prairie potholes typically require standing water to grow, their seeds are only able to germinate on terra firma. So while the brittle mud of a dried-out slough may look dead, it is actually a green-and-growing marsh just waiting to happen. A small square of exposed slough bottom (a square yard, or meter, on a side, dug to the depth of your thumb) can be expected to contain somewhere between 20,000 and 45,000 seeds for as many as two dozen species of plants. The exact nature of this seedbank depends on the history of the pond and the species that set seed in it during the preceding wet-dry cycle. Some of them—like the barbed fruits of nodding beggar-ticks—have been lying there since the last time the slough dried out, anywhere from one to thirty years earlier. As a mud-flat specialist and an annual, the nodding beggar-tick germinates at drawdown, produces a profusion of yellow blooms, and sets and deposits a stock of fresh seed. When water returns to the pond, the mature plants are drowned out, but the seeds of the future lie waiting in the ground.

For cattails and other emergent plants, by contrast, this same inrush of returning water is the stimulus for vigorous growth. Seedlings that germinated on the dry mud flat now sink their roots into the gooey muck, pushing out runners in every direction and sending up juicy young shoots. Then, once the marsh is fully reflooded, pondweed and other submergents begin to germinate in their turn, as the marsh flows into its regeneration phase. Meanwhile, various forms of animal life are also emerging from the mud, in a kind of ritual reenactment of creation. The first to crawl out of the ooze are water fleas and other tiny crustaceans that hatch from eggs, laid the last time the pond was wet, that have remained buried and dormant in the sediments. Some species are so deeply attuned to the wet-dry cycle that they cannot reproduce unless their eggs are first desiccated and then reflooded. This is also true of the "floodwater mosquitoes" (genus *Aedes*), which lay their eggs on plant stalks in dried-out meadows, leave them high and dry all winter long, and rely on spring flooding to revive them. The transition from dormant egg to flying, blood-sucking adult can be achieved in a mere ninety-six hours. Instant mosquitoes: Just add water.

Once the water begins to flood into the pond, life returns in a rush, as these pioneering species soak up the nutrients deposited during the preceding degeneration phase, or die-off. How long this surge of life persists depends,

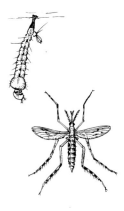

Floodwater mosquito

more than anything else, on how long water remains in the basin. If the slough only stays wet for a matter of weeks, the recovery is cut short and is limited to species that can hatch, mature, and deposit new eggs on a tight schedule. But if the wetland is seasonal or permanent—if it lasts for several months—the recovery is enriched by a wave of colonists. These are mobile species of insects, reptiles, amphibians, mammals, and birds that escaped from the drawdown by moving to deeper, more permanent sloughs and that begin to find their way back as the habitat improves. For example, many kinds of water bugs and beetles cope with the dying-and-reviving nature of wetlands by making regular reconnaissance flights, scanning the landscape for reflective surfaces of an appropriate tint and size. By responding to the glint of water, they can home in on regenerating habitat or, if their instincts happen to fail them, occasionally on the roofs of blue Datsuns! Sometimes they bring stowaways with them, in the form of aquatic mites and crustaceans that have tucked themselves under a wing and that join in the reconstitution of the marsh community. Ducks and other waterfowl also provide transportation from pond to pond, carrying in algae, snails, and other undeclared cargo. (Of the two dozen species of snails found in prairie potholes, several can survive drawdown by burrowing into the bottom or sealing themselves inside a mucilaginous coating. Most, however, can't cope with dry conditions and have to be reimported with each successive cycle.)

With persistent inundation, the pioneering species—the ones that require drawdown to reproduce—gradually fade from the scene and are replaced by organisms adapted to long-term inundation. This category includes, for example, the larvae of tiger salamanders, large, slow-growing, tadpolelike creatures with a collar of fringed gills, which typically spend up to two years in the water before they metamorphose into gill-less, terrestrial adults. Indeed, some individuals avoid making this transition altogether and spend their entire lives in the pond, eventually reaching sexual maturity and even breeding in the form of oversized larvae, called paedomorphs. (Tiger salamanders frequently live to an age of eight to ten years on land, though little is known about the life span of the aquatic forms.)

Tiger salamander
larvae

Yet the high times can't last forever. Inevitably, whether in a matter of months or years, the conditions of life in the pothole will alter and the marsh will cycle on, passing from full flood to degeneration and from drawdown to rebirth.

Inside the Duck Factory

Tiger salamanders spend most of their adult lives in damp crevices and underground dens, only emerging briefly when it rains.

The wet-dry cycle provides a key to understanding how the Prairie Pothole Region is able to produce tens of millions of ducklings in an average growing season. The surge of life that follows drawdown, and the cycling of resources through the marsh, produces a periodic, self-renewing, superabundant crop of algae, minicrustaceans, insect larvae, and other high-protein food that is critical to the success of breeding waterfowl. Even species such as mallards and northern pintails, which are seed eaters for most of the year, switch to a diet of

> ## DRY OUT AND DIE

As the closest living descendants of the ancient, stout-limbed fish that first clambered onto dry land, amphibians retain a strong, lifelong allegiance to water. To paraphrase an old saying, you can take the frog out of the pond, but you cannot take the pond out of the frog.

All the amphibians on the Great Plains, including familiar species of salamanders, frogs, and toads, begin their lives in the water. Hatched from masses of jellied eggs (think eyeballs in aspic) stuck to submerged plants, they develop first into swimming larvae, or tadpoles, that breathe through gills. Then, after an aquatic stage that may last weeks, for frogs and toads, or years, for salamanders, they undergo metamorphosis and are spontaneously transformed into walking, hopping, terrestrial forms.

Yet even as landlubbing adults, most amphibians never dare venture far from a ready supply of moisture. Weak of lung, they supplement their supply of air by breathing through their skins, something that can only happen if their bodies are moistened. If they dry out, they die. This phenomenon explains why the tiger salamander, for example, spends the daylight hours hunkered down in a humid burrow, from which it emerges only at night or during showers. Species such as the northern leopard frog, by comparison, which are highly intolerant of drought, tend to focus their activities on the marshy fringes of streams, ponds, and other permanent sources of water.

There are, however, two species that are especially adapted to cope with arid conditions. They are the Great Plains toad—all wrinkled and warty and blotched—and the plains spadefoot toad, which has smooth skin and looks something like a bulky frog. Found in both desert and grassland environments from the southern part of the Prairie provinces to northern Mexico, they respond to drought by digging a shallow well and burying themselves in the bottom. Both species have spurs, or spades, on their hind feet, which they use to spiral down, rump first, to as much as a yard, or meter, beneath the surface. There they lie, their metabolisms set to idle, for weeks or months or years, until they are reawakened by an influx of rainwater. Then, suddenly, the toads are back on the surface, and the still-refilling rainwater ponds are alive with the quacklike grunts of spadefoots or the metallic, machine-gun-like clatter of Great Plains toads. Eggs are laid and within forty-eight hours develop into tadpoles, which mature into full-grown toads a mere three to six weeks later. Their reproductive cycle completed, the adults feed on insects and other small prey until the drying heat of summer forces them to bury themselves away.

Great Plains toad

Plains spadefoot toad

Northern shovelers

Common mergansers

aquatic invertebrates during the nesting season. It is no small thing, after all, for a female duck to produce a clutch of nine or ten eggs, one a day for as many days, and then to start all over again if the first clutch fails to hatch. (Nest failures are common, and female mallards have been known to make as many as four nesting attempts in a single season.) Because regeneration is usually triggered by snowmelt in the spring, the pulse in the production of pond life is ideally timed to support egg laying and to sustain the hen through the month-long vigil of incubation. And there is still plenty to eat in midseason, when the ducklings peck their way out of their shells, follow their mother to a slough, and bob along after her in search of food.

What's good for ducks is also good for the people who admire them. The delight of traveling through pothole country, especially in a wet spring, derives not only from the numbers of ducks but from their variety. If one pond sports a pair of mallards, the next one will surprise us with blue-winged teals or American wigeons or lesser scaups, each appropriately dapper or dappled or iridescent. Here's a pair of northern shovelers, slowly swaying their big, trowel-like beaks through the shallows, with much the same motion as if they were wielding metal detectors. (In fact, the birds are feeding on microscopic pond life, which they obtain by passing water over their tongues and out the sides of their bills, which are equipped with comblike filters.) In the next pond down the road, a pair of common mergansers may be swimming side by side, he resplendent in green, she with a tousled headdress of cinnamon, their beaks unexpectedly pointed and startlingly red. Hooked at the ends and serrated along both sides, these weapons permit mergansers to catch amphibians and (where the water is sufficiently oxygenated) minnows and other small fishes. No two potholes are likely to yield an identical list of species.

The preference of particular species of ducks for particular potholes is a sign of the subtle-but-significant differences between basins. Through the intricacies of the wet-dry cycle, slight variations in depth and water duration are translated into tangible differences in the amount and types of invertebrates, or duck food, in each wetland. When this slough-to-slough variability is multiplied by millions of basins and extended across the northern plains, it results in a varied landscape of wetland habitats that, collectively, are suited to meet the needs of a varied array of duck species. Pair by pair, the birds select the potholes that are best able to satisfy their specific requirements. Alternatively, if they can't find one wetland that will do the entire job, the birds move from slough to slough, perhaps using one basin for feeding in early spring, another for nesting, and still another for rearing their ducklings. This ability to provide a variety of quality habitat is another reason that the Prairie Pothole Region is such a hotbed of duck production.

Not surprisingly, ducks are more searching in their assessment of wetlands than human observers are, and their reasons for preferring one slough over another are not always apparent to us. Nonetheless, some general patterns are fairly easy to discern. Most ducks that nest on the prairies fall into one of two main groups, or tribes, each of which is adapted to a particular lifestyle. On the one hand are the dabblers or, as they are more formally known, the members of the genus *Anas*, a lineage that includes the mallards, teals, wigeons, gadwalls, pintails, and shovelers. Often seen upended, tail waggling, in the slough, these ducks are adapted either to feed near the surface of the water (like shovelers) or, by tipping up, to forage on the slough bottom. Accordingly, they prefer shallow potholes, usually with an interspersion of open water and emergent plants. Since their preferred habitat is highly dynamic—flooded one year, dry the next—they tend to be opportunistic in their choice of breeding grounds, with each pair settling in the first unclaimed slough that they happen to come upon during the spring migration. Finally, because their chosen potholes are not deep enough to offer protection from terrestrial predators, dabbling ducks often nest in dense vegetation well away from wetlands and lead their ducklings on a long trek back to water.

The other tribe of pothole-nesting ducks is the divers. Less abundant and perhaps less familiar than the dabblers, this group includes two eye-teasing pairs of lookalikes—the canvasback and redhead, and the ring-necked duck and lesser scaup—all of them members of the genus *Aythya*. An unusual assortment of other, less closely related species can also conveniently be mentioned in the same breath, notably, common goldeneyes, buffleheads, mergansers, and ruddy ducks. (Ruddy ducks, in particular, are worth checking out, if only for the bright blue bills of the breeding males.) What these birds all have in common is their ability to dive, swim underwater, and exploit the resources of the deeper and more permanent prairie marshes. Unlike dabblers, which prefer ephemeral and seasonal basins, divers tend to be found in seasonal and permanent potholes with areas of open water. Although several of the diving species protect their eggs by nesting in hollow trees, others find safety for their broods on islands or in stands of emergent plants in the middle of the basin. And since their chosen waters are relatively reliable, as prairie potholes go, diving ducks tend to return to the same nesting sites year after year, rather than playing catch-as-catch-can in the way that dabbling ducks do.

Reliable, as prairie potholes go. The fact is, of course, that even the deepest and most permanent wetlands in the pothole region cannot be depended upon. A succession of snow-free winters and hot summers can dry the whole region out, including even the most extensive and durable marshes. Clearly, both dabbling and diving species are adapted to thrive in these dynamic environments when conditions are "ducky," but what happens when they're not? How do waterfowl cope when their habitat dries up?

The short answer is, they don't. Instead, like other types of pond life, ducks have little choice except to get by as best they can. Most species of pothole-nesting ducks tend to overfly the dried-out potholes and continue on to the north, often settling down in the boreal forest. In this less-than-ideal environment, they produce a meager crop of young, often too few to sustain the breeding population. Thus, fewer sloughs translates directly into fewer ducks. This relationship between the number of ponds in the pothole region in May and the number of breeding ducks tallied the following spring has been documented repeatedly in data that go back to the early 1960s in Canada and the mid-1970s in the United States. Although the graphs do not line up precisely, because of the many complicating variables that influence the ducks' success (including the timing of spring breakup, unseasonable blizzards, and so on), the parallel run of the lines from peak to trough is too striking to be questioned.

Where Have All the Ducklings Gone?

The bleakest years so far on record for North American ducks extended from the mid-1980s to the early 1990s, a period during which the climate wavered queasily between wet and dry for several years before settling into the most intense drought since the Dirty Thirties. The dabbling ducks, in particular, were hit hard by this turn of events, as durable species like the mallard plunged toward the lowest numbers ever recorded. (In 1985, this most populous of North American ducks bottomed out at about 5 million birds, a worrying 40 percent below its long-term average numbers.) Worse yet, even during the occasional seasons when the prairie potholes did fill up, the ducks inexplicably failed to rebound. It seemed that some unknown difficulty was compounding the effects of the drought and preventing the ducks from recovering during the wet phase of the cycle. Alarmed that this downturn might be permanent, researchers embarked on an urgent search for answers.

One fact was immediately obvious. The Prairie Pothole Region was no longer the paradise for ducks that it had been at the beginning of the settlement era. Since 1900, somewhere between 20 and 85 percent of the original prairie

Already half-submerged, this eared grebe could slip underwater in an instant. These handsome birds often breed in large, noisy colonies on marshy, freshwater lakes where they dive to catch aquatic insects.
Arthur Savage photo

Garter snakes are commonly found near wetlands and river valleys, on the prowl for frogs, earthworms, and small mammals.

potholes have been ditched, drained, dredged, filled, leveled, plowed, or paved; the exact percentage varies from district to district. (In general, losses have been lowest where land is used for grazing and highest on croplands and in urban settings.) Although the rate of wetland conversion has declined in recent decades—and the pace of restoration has picked up—the net balance is still negative. Between the late 1970s and the early 1980s, for example, an additional 1,500 square miles (4,000 square kilometers) of wetlands were appropriated for farming operations in the U.S. alone, and the losses in Canada were similar. This was the very period when the continental population of ducks began to slide into the basement.

Loss of wetlands is never good news for ducks, yet biologists were not convinced that it was a primary cause of the 1980s crisis. After all, the immediate problem wasn't really a lack of potholes; it was the inability of the ducks to ride the wave of recovery following drawdown and fully reoccupy the wetlands still available to them. Some link in the chain between the wet-dry cycle and duck reproduction had been broken, so an abundance of duck food during wet

> QUEER DUCKS

In his *Life Histories of Familiar North American Birds,* pioneering ornithologist Arthur Cleveland Bent described an unusual sighting of a northern pintail. "Once," he wrote, "while crossing a tract of burned prairie [in the early 1900s], I saw a dark object fully half a mile away, which on closer inspection proved to be a pintail sitting on a nest full of half roasted eggs; this was a beautiful illustration of parental devotion and showed that the bird was not dependent on concealment."

The northern pintail is quite literally a queer duck. A member of the dabbling duck tribe, it looks a bit like a mallard that has been stretched at both ends to produce an elongated neck and, in the male, an extravagant, quill-like tail. The drakes are also distinguished by a sleek chocolate-brown head, a pure white breast, and a swirled boundary where the colors meet on the neck. (If an Art Deco designer had set out to create a duck, it might well have resembled a male pintail.) Hens, are clad in camouflage plumage the color of dry grass, plain but elegant.

What makes northern pintails unusual is their lackadaisical approach to the selection of nesting sites. Unlike most other dabblers, which conceal their nests in thickets of grass or shrubs, pintails settle down wherever the spirit moves them, even in sites that are completely open. Under natural conditions, when their nests were mere points in unbroken vistas of grass, this strategy presumably made some kind of sense. A predator had to be lucky to find them. But these days, when the grasslands are broken into fields and fragments and liberally studded with trees, hawks and other predators have gained a troubling advantage.

Pintails that nest in crop stubble are at special risk. A recent study of nest success in an intensively farmed area of southern Saskatchewan determined that out of every hundred stubble-field nests, no more than four were successful. The reasons for this abysmal record were spring seeding, which destroyed 20 to 33 percent of the nests, and high levels of predation, which took out the rest. (Nests in denser cover enjoyed greater success, at 6 to 37 percent.) As a result, pintails are the only species of predominantly pothole-nesting ducks that have not rebounded from the 1980s drought. "Parental devotion" has not been enough. But the people at Ducks Unlimited hope that a shift in agricultural practices toward the planting of fall-seeded, or winter, wheat may give pintails a better chance by reducing spring disturbance and providing early-season cover for nests.

Northern pintails

years no longer resulted in an abundant output of ducklings. Could it be, the researchers began to wonder, that something was suppressing duck reproduction? Studies conducted in the early-to-mid-1980s quickly confirmed this disturbing hypothesis. Instead of the 15 to 20 percent "hatchability" (one or more hatchling per nest) that is considered a bare minimum to keep duck populations at a steady state, the results from the Prairie Pothole Region were dismal: an average of 7 percent nest success for pintails, 12 percent for shovelers, 14 percent for gadwalls, and 8 and 11 percent for mallards in two different areas. These figures meant that even when the ponds were crowded with ducks and the nests were brimful of eggs, the vast majority of nesting attempts were failures. The famed duck factory of North America appeared to have gone bust, and a once-productive population was now merely going through the motions of reproduction.

One curious detail—a needle in the haystack of data—emerged from this research. The studies showed a strong and surprising relationship between the prospects of a given nest and the human land use in the countryside that surrounded it. Specifically, the greater the percentage of cropland in the area, the lower the likelihood that the ducks would be successful. This was partly because spring-seeding activities took a toll on nests that were situated in stubble fields, sites that are especially favored by pintails. But this kind of collateral damage did not appear to be a significant problem for other species of dabbling ducks, which nest in tractor-no-go zones such as ditches and patches of brush. The significant problem appeared to be not farming as such but one of its unintended consequences. In one way or another, agriculture had inadvertently boosted the populations of middle-sized predators, such as foxes, raccoons, and skunks. By the mid-1980s, this cadre of nest-raiding mammals (aided by bullsnakes in some localities) was making a meal out of both duck eggs and sitting ducks, and there didn't seem to be much that anyone could do to stop them.

Then two apparently unrelated events occurred. In 1985, as drought took hold of the prairies in earnest, the U.S. government introduced a program designed to prevent soil erosion and support the region's beleaguered farmers. Known as the Conservation Reserve Program, or CRP, this initiative provided an annual payment, on a 10- or 15-year term, to landowners who agreed to retire unproductive land and plant it to grass, alfalfa, or other forms of soil-binding permanent cover. Conceived as a temporary response to hard times, the CRP was subsequently renewed, updated, and expanded to buy protection for thousands of fragments of upland habitat in the Dakotas and adjacent parts of the

Pothole Region. In so doing, the program almost inadvertently created a vast new resource of quality upland habitat for wild animals and birds. With better cover for their nests, ducks on CRP lands frequently enjoyed success rates in the range of 20 to 30 percent, more than enough to lift the populations out of their slump. (Despite this quacking dividend, the Conservation Reserve Program has recently been curtailed, and millions of acres are expended to be returned to cultivation in the next few years.)

Meanwhile, at the same time as new habitat was being created, ducks received another boost when a decade of drought finally drew to a close. When the long-awaited rains returned in the early 1990s and the sloughs filled up, the Prairie Pothole Region moved from drawdown to regeneration in a headlong rush. After a period when it had seemed that ducks were on their way down, they were suddenly present in every ditch and slough, and the populations of several species surged toward record numbers. For the first time in history, however, the pulse that replenished the population came mainly from the CRP lands in the Dakotas, rather than from the Canadian prairies, which had been the powerhouse of duck production in previous surveys. This development was widely seen as a warning that the recovery was precarious, especially in districts where large tracts of land were under cultivation. Yet for all the stresses and strains of the previous century, the ability of pothole-nesting ducks to rebound from near-disaster stands as a testament to the regenerative powers of prairie wetlands and the seemingly inexhaustible resilience of the wet-dry cycle.

Prairie Rivers

Prairie potholes and other ephemeral wetlands are classic examples of ecosystems adapted to cope with almost constant disturbance. Faced with a relentlessly dynamic environment, pond life has become almost as fluid and responsive as water itself. And this same description also applies, though in strikingly different ways, to the life of the two major river systems that cut across the Great Plains. They are, to the north, the Saskatchewan and, to the south, the Missouri.

The Saskatchewan system extends across the Canadian plains like a spindly arm, with its fingertips in the icefields of the Alberta Rockies and its elbow at the north end of Lake Winnipeg. Rising as a thousand milky meltwater trickles that tumble down the continental divide, the water converges into creeks and then rivers that flow to the north and west and ultimately join to form the

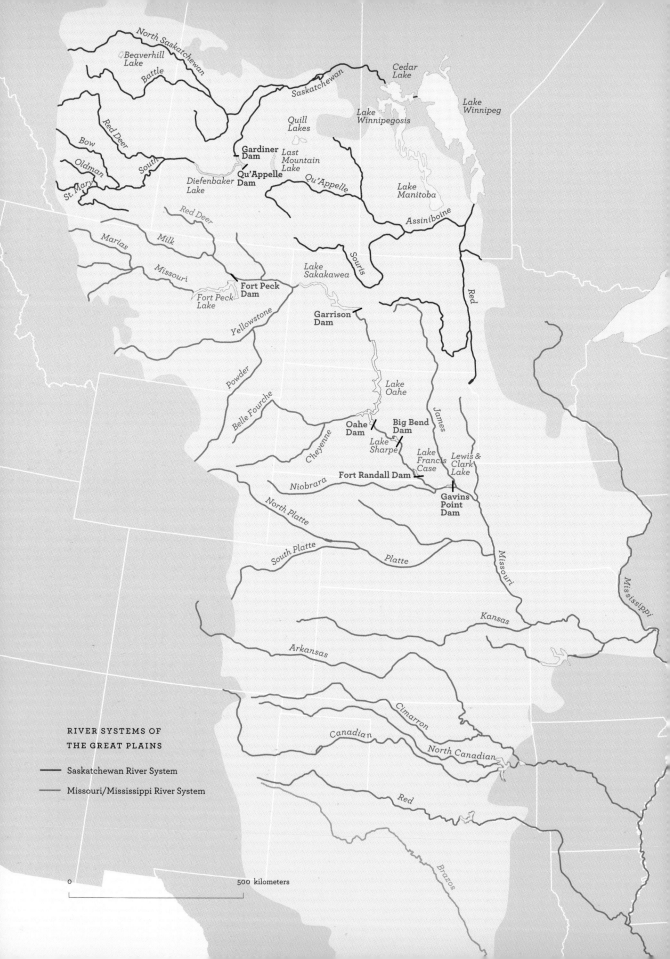

North Saskatchewan

Beaverhill
Lake

Battle

Cedar
Lake

Saskatchewan

Lake Winnipeg

Lake
Winnipegosis

Quill
Lakes

Red Deer

Bow

South

Gardiner
Dam

Last
Mountain
Lake

Qu'Appelle
Dam

Diefenbaker
Lake

Qu'Appelle

Lake
Manitoba

Oldman

St. Mary

Assiniboine

Red Deer

Marias

Milk

Souris

Red

Missouri

Lake
Sakakawea

Fort Peck
Dam

Fort Peck
Lake

Garrison
Dam

Yellowstone

Powder

Lake
Oahe

James

Belle Fourche

Oahe
Dam

Big Bend
Dam

Cheyenne

Lake
Sharpe

Lewis &
Clark
Lake

Lake
Francis
Case

Niobrara

Fort Randall Dam

Gavins
Point
Dam

North Platte

South Platte

Platte

Missouri

Kansas

Arkansas

Mississippi

Cimarron

Canadian

North Canadian

RIVER SYSTEMS OF
THE GREAT PLAINS

——— Saskatchewan River System

——— Missouri/Mississippi River System

Red

Brazos

0   500 kilometers

North Saskatchewan River. Meanwhile, water cascades down the mountains to the south and west into the valleys of the Oldman, Bow, and Red Deer rivers; together, they become the lovely, often lazy South Saskatchewan River. Cutting cross-country through the farmlands of Alberta and Saskatchewan, the two branches meet east of Prince Albert to form a single, gray-green flow, the Saskatchewan River. From there, it is on to Lake Winnipeg and then north, via the Nelson River, into the Arctic Ocean at Hudson Bay. From headwaters to mouth, the major rivers of this system cover a combined distance of some 1,600 miles (2,600 kilometers) and provide drainage to 400,000 square miles (or more than a million square kilometers) of land in the three Prairie provinces.

The Missouri River system is, if anything, even more magnificent. From sources in the Bitterroot Range of northwestern Wyoming and southwestern Montana, the mainstream of the Missouri meanders east across Montana and then south through the Dakotas before emptying its muddy flow into the Mississippi River at St. Louis, for a total run of almost 2,500 miles (4,000 kilometers). Routed along the shore of a long-vanished Ice Age glacier, the great river snakes diagonally across the center of the continent, attracting drainage from both the north (through tributaries such as the Milk and the Frenchman) and the west (through, among others, the Yellowstone, Little Missouri, Niobrara, and Platte rivers). All told, the rivers of the Missouri system drain an area of about 530,000 square miles (1.4 million square kilometers), including the southernmost Canadian prairies and one-sixth of the continental United States. That's the entire sweep of country from the Rockies to northeastern Missouri and from southern Saskatchewan to central Kansas.

From central Kansas south, by contrast, the land slopes ever so gradually away from the Missouri basin and entrains the rivers to run directly into the Mississippi River or the Gulf of Mexico. These include the Cimarron, Canadian, and Arkansas rivers of the Arkansas River system, and the Brazos River of central Texas. Thus, in one way or another, the waters of the dry-and-dusty American heartland all end up in the Caribbean Sea and, thence, in the North Atlantic.

Of Fish and Flood

Although we tend to think of rivers as if they were all alike—just an onward rush of liquid confined between banks—every watercourse is unique, distinguished by a particular set of physical characteristics. Does the water freeze solid in winter or flow freely throughout the year? Is it fast and deep or soupy

and shallow? Clear as crystal or burdened with eroded soil? Does it shrivel away to nothing in midsummer? The answers to these questions define the conditions of life in a particular stream and determine the range of species adapted to survive in it. Thus, rivers like the North and South Saskatchewan and upper Missouri, which are ice-bound for part of the year, support far fewer types of fish (some two or three dozen at most) than rivers farther south, like the Arkansas and lower Missouri (each with a species list of 100-plus). Similarly, the seasonally low and often oxygen-deficient waters of the arid western plains tend to have fewer types of fish than the more generous and stable rivers of the tall-grass prairies. In the eastward-flowing Arkansas River, for example, the species count more than doubles from west to east, rising from 54 to 117.

A river is not really one place but many: a long, sinuous linkage of different environments connected by the flowing fabric of the water. Even adjacent stretches of the same river can be remarkably dissimilar, from sparkling, well-oxygenated riffles where the water slips over gravel beds to slow, swirling pools

behind logjams to marshy backwaters set away from the main channel. Here fish spawn or feed or seek refuge from predators. There frogs lay their eggs, turtles bask in the sun, and great blue herons stalk through the shallows. The diversity of life in the river depends on the ability of the water and its basin to maintain a constant supply, or flow, of different habitats, so that a diversity of new opportunities is continually created.

The temperature shock of winter is just one of the variables that define the character of a stream or river.

Arthur Savage photo

For the major rivers in both the Saskatchewan and Missouri systems, the principal force of renewal has, until quite recently at least, been the chaotic ritual of spring flooding. As so-called large floodplain rivers, the North and South Saskatchewan and the Missouri are naturally characterized by highly dynamic flows and were never meant to be confined within narrow channels. In the case of the Missouri, to cite the best-documented example, the river followed a seasonal rhythm, falling to its lowest levels in winter, experiencing a small surge at breakup in spring, and then, depending on the year, often really kicking over the traces in early June. Forced by the onrush of snowmelt from the mountains, the

Painted turtle

The original dam builders on the prairies were not men equipped with bulldozers and engineering degrees but a brigade of fat, fur-bearing mammals armed with sharp front teeth. They were beavers, surely the only rodents in the world to have provided motivation for the expansion of empire.

The gleam that first drew Euro-Americans onto the plains was the sheen of the beaver's sleek fur, a material that worked up handsomely into fashionable, felted apparel. During the two and a half centuries of the fur trade, the demand for pelts was so intense that an initial continentwide population of some 60 million to 400 million beavers was reduced to near extinction. Since the 1930s, however, when felted-beaver top hats were no longer à la mode, the numbers have rebounded to a respectable 6 million to 12 million animals.

This is good news not only for beavers and their human admirers but also for the diverse community of plants, insects, amphibians, reptiles, mammals, and birds that rely on beaver-created habitats. By damming creeks and rivers, beavers create ponds that are both quiet enough to permit them to stockpile food for the winter—dark, submerged tangles of aspen branches and trunks—and deep enough to retain free water after freezeup. Thus, beavers can slip from their underwater lodges to their underwater caches to feed without venturing out into the Great Deep Freeze.

By rearranging the world to suit themselves, beavers inadvertently create benefits for other creatures. As the water collects behind a dam, it spills out over the banks to flood a broad margin of shallow wetlands. Still connected to the main current of the river, this network of meadows and marshes provides habitat not only for aquatic plants but also for spawning fish, frogs, salamanders, dragonfly larvae, mosquito wrigglers, garter snakes, muskrats, and a whole range of other creatures that are not commonly found in fast-running rivers.

The cost of these ecological enhancements is paid primarily by the local aspen. Although the trees grow back vigorously as the beavers gnaw them off, the demands are often so heavy that the aspen can't keep up. After a decade or two of high-powered eating, the beaver are generally forced to move to an unexploited stretch of the river, where they can start anew. Meanwhile, at the old dam site, the abandoned pond—now fringed with regenerating stands of aspen—begins to fill up with silt and plants, changing from pond to marsh to meadow in an ever-shifting mosaic of habitats.

Northern pike

river swirled up over its banks and out across its wide, flat-bottomed overflow valley, or floodplain. With the full force of a natural disaster, the water scoured away sandbars, uprooted shrubbery and full-grown trees, tore out logjams, and generally ran amuck, leaving the entire floodplain in chaos.

But all that mess and disorder held the seeds of rebirth. When the flood subsided, it revealed a soggy landscape of ponds and puddles, many of them linked to the river's main flow, that would quickly grow up into a lush and varied mosaic of oxbow lakes, marshes, and wet meadows. Amphibians laid their eggs in these quiet waters; certain fish, like the northern pike, splashed through the

shallows to spawn; others relied on the floodplains as rearing grounds for their young. And all this productivity was accelerated by an input of nutrients from the rotting muck that was strewn around the countryside in the wake of the flood. In this way, a landscape of aging floodplain forests and silted-up wetlands was periodically restored to the freshness and vigor of youth, and the resources of the system were freed up to support a new burst of growth.

Meanwhile, as the main flow of the river fell back into its channel (around many a new twist and bend), its waters were muddied with silt and decaying vegetation. Yet this apparent burden constituted a food-rich resource for the community of plankton in the river and hence for plankton eaters, such as insect larvae, crustaceans, and fish, and ultimately for most other organisms that lived in the stream. This surge of productivity is known as the flood-pulse, and, though still poorly understood, it appears to be as important to the life of rivers as the wet-dry cycle is to marshes. For example, many species of river fish—including native prairie minnows like the Arkansas River shiner and speckled chub—rely on spring flooding as a signal to spawn. By responding to

During the winter, the northern leopard frog often hibernates on the muddy bottom of a creek, river, or lake— virtually motionless— then pops back up to the surface when spring comes. Like many other species of frogs around the world, the northern leopard frog has suffered recent, severe declines in population, probably the combined result of habitat loss, disease, pollution, and climate change.

Johnny darter

Green sunfish

facing page: Often found in swampy margins along rivers and creeks, the great blue heron uses its bill not only for preening but also for nabbing slippery frogs and fish.

this seasonal cue, the fish are able to coordinate their life cycle with the turning of the year, thereby ensuring that their young receive the full benefit of both the flood-pulse of productivity and the growth-enhancing warmth of midsummer.

In the last half century, however, this natural rhythm has been disturbed. In some instances, the seasonal flood cycle has been dampened by an overall lowering of the rivers' flow caused by heavy withdrawals for irrigation. (These demands have intensified since World War II, when pumps powered by surplus airplane engines were first hooked up to central-pivot irrigation systems.) As water is pumped out of underground formations, the level of the groundwater falls, just as if it were being sucked out by thousands of high-tech drinking straws. As the water table drops, there are no reserves left near the surface to trickle into the river, through seeps and springs, and help to keep its flow replenished. As a result, the river is now so diminished that it can no longer overflow, flood the surrounding plain, and replenish the groundwater through a trickle-down effect. Caught in this self-defeating cycle, the water table underlying much of the southern plains dropped by anywhere from 10 to more than 100 feet (3 to more than 30 meters) between 1940 and 1981, and the flow of many rivers was also drastically curtailed. To cite one poignant example, the Kansas stretches of the Arkansas River have dwindled from an average flow of 49 cubic feet per second, pre-1960, to a mere 7 cubic feet per second in 2005, a decrease of 86 percent.

Not surprisingly, this degradation has taken a heavy toll on species that are adapted to live in free-flowing, seasonally flooding rivers. A century or so ago, that description fit most rivers on the southern plains, including the

little Smoky Hill River of west-central Kansas. When the fish of the Smoky Hill basin were first catalogued in 1885, researchers hauled in specimens by the dozens. The list of species was evocative: hornyhead chubs, johnnydarters, brassy and suckermouth minnows, black bullheads, stonecats, green sunfish, blacknose shiners, southern redbelly dace, and many others. In 1985, when the survey was repeated, all the above-named species were gone, a change attributed to reduced water levels, loss of habitats, and the suppression of the flood-pulse cycle. In their place were a relatively small number of more-or-less ubiquitous introduced fishes, including largemouth bass and carp, that were able to cope with the new conditions.

The withdrawal of groundwater for irrigation is not the only postwar project that has left its mark on the ecological integrity of prairie rivers. Since the 1950s, engineers have also been busy constructing dams, diversions, impoundments, and other impediments to ensure that water is made available for human purposes. Across the entire Great Plains, with its thirsty cities and farms, there is scarcely a stream of any significance that has not suffered some such "improvement." For instance, the North Saskatchewan River is held back by two major impoundments before it even reaches the plains, pooling out to form Abraham Lake (in the mountains west of Red Deer) and the Brazeau Reservoir (southwest of Edmonton). The South Saskatchewan River is backed up by the Gardiner Dam (south of Saskatoon) at the head of the long noodlelike contours of Lake Diefenbaker. And this is to say nothing of the dozens of small reservoirs on dozens of tributaries that affect the movement of water through the Saskatchewan River system.

Then there's the Missouri. Now more accurately described as a string of reservoirs than as a river, the upper Missouri is blocked by seven major dams, all but one of them built in the immediate postwar period. Working our way downstream from the river's headwaters, they are: Canyon Ferry (1952) and Fort Peck dams (1940) in Montana; Garrison (1953) in North Dakota; Oahe (1958), Big Bend (1963), and Fort Randall (1954) in South Dakota; and Gavins Point (1955), at Yankton, near the South Dakota/Nebraska border. Between Sioux City and St. Louis, by contrast, the lower Missouri has been intensively reengineered for a different use. Instead of being interrupted by dams, this "home stretch" of the river has been dredged, deepened, and banked to convert it into a fast-flowing, uninterrupted channel for barges. Together, these alterations have been admirably effective at meeting their stated goals, whether navigation, power generation, recreation, or flood control. But they have also—to an extent that seems inevitable with hindsight but that was initially unforeseen—drastically altered the life of the Missouri.

The whole idea of damming rivers is to regulate seasonal variations in flow and make a chaotic natural system more predictable. Rather than permit spring meltwaters to rush out over the floodplain—by now a complex of cornfields, multilane freeways, and urban sprawl—engineers can hold back the flood behind massive barriers. With the exception of freak seasons like the spring of 1993 (when rapid snowmelt and heavy rains combined to overwhelm the system and create havoc throughout much of the Missouri basin), the creative tumult of spring flooding is over and done with. And it is not just the magnitude of the peak flows that is now under human control: their natural rhythm has also been altered. Under modern conditions, the volume of water that is permitted to pass through the sluicegates depends not on the logic of the seasons but on human demand. As a result, the more-or-less free-flowing stretches of the Missouri (between and below the dams) now typically experience their peaks not in spring but in winter, when the draw on the system is low, and fall to their lowest ebb during the growing season. This change completely confounds the flood pulse on which so many river-adapted organisms depend and makes the river a less livable environment for them.

Even changes that, at first glance, look like straightforward improvements often net a mixed bag of gains and losses. For instance, the Missouri, like many other prairie rivers, is naturally heavy with mud, the result of a long, slow journey through the dry and erodible soils of the grasslands. (They don't call it the Big Muddy for nothin'.) Also, like other prairie rivers, it is typically shallow—"a mile wide and an inch deep," as the settlers used to describe it. Warm, murky, and sullen, it is the antithesis of the proverbial babbling brook and, to some early observers, appeared to have few natural virtues. "Too thick to drink, too thin to plow," the pioneers grumbled. But if you hold this same water behind a dam, it will clear, as the load of dissolved sediments settles to the bottom. This is what happens in present-day reservoirs, creating perfect conditions—cold and sparkling clean—for sight-feeding game fish such as the native walleye (now far more common and widespread in the Missouri basin than in times past) and the introduced northern pike, smallmouth bass, black crappie, chinook salmon, and others. In all, about half the fish species in prairie rivers these days are exotics, and they tend to be doing very nicely.

But what of the poor unfortunates that liked the turbid water of the Missouri exactly the way it was? What about creatures like the pallid sturgeon? One of approximately two dozen species of native Missouri River fish that are known to be in decline, the pallid sturgeon is among the unlucky few to merit inclusion on the federal list of endangered species. A member of an ancient order of cartilaginous fishes that has survived almost unchanged for some 70 million years, this is one weird-looking creature. It's big (up to 6 feet long and 100 pounds in weight, or 1.8 meters and 45 kilograms) and looks vaguely menacing, with its gray, armor-plated body; small, beady eyes; and flat, shovel-shaped proboscis. Four fleshy whiskers, or barbels, dangle under this protruding snout and finger the water in front of the recessed mouth. These waggling appendages are sensory organs that, through a combination of touch, taste, and electrical cues, permit the sturgeon to locate food in even the siltiest waters. (Pallid sturgeon eat small prey such as insect larvae and worms, which they vacuum up off the river bottom.) Interestingly, researchers in North Dakota have recently discovered similar adaptations in two native prairie minnows—the sicklefin chub and sturgeon chub—which, though weak-eyed, feel their way through the gloom by means of special sensory bumps located on their heads, fins, and bellies and inside their mouths. Thus, there appears to be a guild of prairie fishes that are specially equipped to cope with the special challenges of prairie rivers.

As the character of the habitat changes—as a silty, warm, seasonal river is transformed into a clear, cold, managed flow—these specializations lose their survival value. What is the point of being adapted to heavy siltation or to a seasonal flood-pulse or to a dynamic flow of habitats if many of these natural processes have been altered? This appears to be the fatal question for many native prairie species of fish, including the pallid sturgeon. A survivor of the Ice Age, it has not been able to adapt to the changes brought about by a half century of human ambition. Yet if evolution cannot undo the damage, perhaps people can. Although major dams are unlikely to be removed from the Missouri system any time soon, conservationists are working on ways to modify the obstructions so the fish can regain access to the remaining stretches of free-flowing current.

Even without human remediation, the pallid sturgeon, together with a number of other declining species, such as the western silvery minnow and the flathead, sicklefin, and sturgeon chubs, are managing to hang on in a few restricted reaches of the Missouri River. These include, as a prime example, a

> LAKE STURGEON

Four members of the sturgeon family are found in the rivers and lakes of the Great Plains, and all of them are wonderfully odd. They include the pallid sturgeon (now one of the rarest fish in North America), the shovelnose sturgeon (a smaller, darker version of the pallid, so similar that the two frequently inter-breed), and the paddlefish (perhaps the strangest of the lot, with its long, spatulate snout—a third of its entire length—which picks up electrical signals from the schools of plankton on which it feeds). All of these species are native to the Missouri/Mississippi drainage.

Only the final member of the family has a range that extends beyond the Missouri system and into the Saskatchewan basin. Found in both the North and South Saskatchewan rivers, the lake sturgeon looks more or less like the pallid—the same armored body, beady eyes, and whiskered chin—but can be considerably larger. The biggest lake sturgeons on record reached lengths of 8 feet (2.5 meters) and topped the scales at 275 pounds (125 kilograms). Gentle giants that they are, they subsist on a diet of fish eggs, insect larvae, and other small organisms that they scoop from the bottom. Slow growing and slow to mature, they have been known to live for up to 150 years. Because of heavy commercial fishing and environmental damage caused by dams, the populations have declined drastically in recent decades, with reductions in the order of 50 to 80 percent since 1980, and only restricted sport fishing is permitted.

In his book *Fishing in the West,* writer David Carpenter has this to say about angling for lake sturgeon:

The standard method of catching sturgeon is to bring a fishing rod in one hand and a good book in the other to the edge of your favourite river. You cast your bait out to a likely spot, let it sink to the bottom and wait. Most anglers lay their rod in the crotch of a forked stick plunged into the beach, and some place a little bell on the tip of their rod to wake them if they've chosen the wrong book. (I would recommend *Moby Dick.*) . . .

It takes at least [twenty-five to thirty years] before these fish can spawn, so they are easily depleted. It behooves us to keep an eye on the populations of these leathery giants.

Lake sturgeon

Pallid sturgeon

> COOL AS A CLAM

Just when you thought you had heard everything, someone tells you about the "unionid mussels." Turns out they're aquatic mollusks, or shellfish, that live in creeks and rivers across the Great Plains (among other places) and that produce the pearl-lined, ridged-backed, more-or-less-oval shells sometimes to be found around muskrat lodges or strewn on beaches. Often referred to as "freshwater clams," they are an important source of food not only for muskrats but also for raccoons, mink, otters, fishes, and certain birds. Humans can eat them, too, but only after they recover from the initial state of shock that inevitably follows from making their acquaintance.

At first glance, the life of a mussel sounds like boredom in the extreme. This is an animal that lies on the river bottom, half buried in gravel or sand, and pumps water in one end of its body and out the other. In the process, it filters out particles of food, mostly plankton or other bits of organic matter. And it goes on doing this continuously through a life span that, depending on the species, can easily extend from 10 to 100 years.

But things start to get interesting when the mussels breed. The first thing that happens is that the male mussel (yes, there are two sexes) releases clouds of sperm into the water. When a female sucks in this seminal fluid, the sperm fertilize her eggs, which develop into larvae inside her body and attach themselves to her gills. To complete their development, however, the larvae have to abandon mother

and launch themselves into the world by hitching a ride on a fish. Although some species of mussels can latch onto any available transport, others are specialists that can only accept the services of particular hosts.

How to make contact? In some species, the female simply turns the larvae loose and lets nature take its course. Either they find suitable hosts or they don't. But many mussels are able to reduce the risk of failure by luring the fish in close. The female plain pocketbook mussel, for example, has a lip of tissue, or mantle flap, that looks astonishingly like a small shiner, complete with eyes and stripes. (Other species present lures that resemble crayfish and insect larvae.) By contracting this appendage, the mussel can wiggle it to draw in nearby bass, its host, and trick the fish into striking. Instead of a meal, the bass ends up with a mouthful of mussel larvae, which hook onto its gills or fins and grow to maturity. Once fully developed (usually without harm to the fish), the young mussels drop to the river bottom, where if conditions chance to be right, they settle into the humdrum routine of adult life.

Because they are highly sensitive to changes in their environment, whether through soil erosion, siltation, dredging, channelization, impoundments, or the expansion of exotic species such as zebra mussels, these wonderfully bizarre creatures are among the most endangered groups on the Great Plains and elsewhere in North America.

short run of water just east of the Montana/North Dakota border and west of Lake Sakakawea, near the confluence of the Missouri and the Yellowstone rivers. The last major river on the Great Plains without a major dam, the Yellowstone not only maintains its natural vitality—with May and June rises, fast- and slow-water habitats, natural temperatures and turbidity, and shifting banks—but also possesses enough energy to reinvigorate the smaller and more docile Missouri. The fact that the native community of fish respond to this wild influence, however locally, suggests that restoring sections of the river is a credible strategy. One of the places in which this kind of restoration is being attempted is the Big Muddy National Fish and Wildlife Refuge in the state of Missouri. From the early 1990s onward, stretches of the river have been released from the demands of navigation and allowed to flood their bottomlands, thereby creating new sandbars and side channels. In 1998, these efforts were rewarded by the discovery of baby pallid sturgeon, no larger than tadpoles, in one of the restored backwaters. The first sign of natural reproduction by the species in the lower Missouri in at least fifty years, these bewhiskered little creatures were a welcome reminder of the synergy between life and the living waters of the Great Plains. Tragically, there has been no natural reproduction in the Upper Missouri in living memory.

In late 2009, the U.S. Army Corps of Engineers (the organization that was charged with "developing" the Missouri River all those years ago) embarked on a five-year, $25,000,000 study to reconsider the overall costs and benefits of their efforts. Laying aside their heavy equipment for the moment, they hope to initiate a national conversation about the Missouri's health. In light of what we have learned about the ecological consequences of dams, do the original reasons for controlling the river command respect? This story is marked *To Be Continued.*

seven

PRAIRIE WOODLANDS

And I saw that the sacred hoop of my people was one
of many hoops that made one circle, wide as daylight
and as starlight, and in the center grew one mighty
flowering tree to shelter all the children.

BLACK ELK, *BLACK ELK SPEAKS*, 1932

IF YOU PAINT the prairie with broad strokes, all you see are
grass and sky—a ground of green or khaki, a wash of blue.
But scan the scene more closely, looking for shadow as well
as for light, and you will see that the face of the grassland is
etched with trees and shrubs. Tangles of woody vegetation are
scrawled along the contours of the Great Plains, defining coulees, streambeds, and ridges and generally adding emphasis to
a landscape that suddenly reveals itself in three dimensions.
Far from being characterized by the absence of trees, these so-
called treeless plains are increasingly a land of many small, scattered woods, each a world in itself, that add complexity to the
grasslands.

It takes a prairie dweller to fully appreciate the way trees
change the world. To sense how they cut the wind and scatter
the light. To smell the rising odors of sap and moisture and
rotting earth. To hear the silvery rustle of leaves and the bright
chorus of birds—warblers, thrushes, vireos, wrens—that flit
and flash through the branches. Every prairie grove is a woody
oasis in a sea of grass, a refuge for the hundreds of species of
insects, amphibians, reptiles, mammals, and birds that are
adapted to the shelter of woodlands. Yet as much as we enjoy
this abundance, we seldom stop to think about the woody

beings that stand at the center of things. What are these impressive organisms that we call trees and shrubs? And what are they doing on the grasslands?

To Be a Tree

The underlying principle of woody plants is height. Through evolution, their strategy has been to hoist their leaves up into the light atop either a single, towering trunk (thereby producing the form classified as a tree) or a cluster of often-shorter, multiple stems (a shrub). The "technology" that makes this possible is an elaborately modified sugar, or organic polymer, called lignin. Produced by a layer of cells just under the bark, this remarkable material—durable, springy, and tough—is laid down toward the center of the stem or trunk, where it bonds with other plant tissues to form a supporting core of wood. Remarkably, this adaptation has been achieved independently by many different kinds of plants, including grasses (bamboo), legumes (caragana), and asters (sagebrush), as well as by all the varied tribes of conifers and hardwoods. In every case, the benefit has been the same: to gain an advantage over ground-hugging grasses and forbs by reaching up and intercepting the sunlight in midair.

But every adaptation brings its own difficulties. The closer a plant is to the heavens, the farther it is from Earth, and leaves, no matter how lofty, still must be fed by their roots. Like the photosynthetic organs of all plants large and small, tree leaves need a supply of water and mineral nutrients that can only come from the soil. Woody plants respond to this challenge by drawing groundwater up to the leaves through slender tubes in their phloem, or inner bark. Since the roots cannot push the fluid up, it has to be drawn from the top, a feat accomplished through evaporation. As the sun warms the surface of the leaves, water vapor is drawn through their pores, or stomata, and lost to the atmosphere. This loss creates a partial vacuum in the leaves that causes liquid to rise up the phloem tubes, each molecule glued to the next, as if it were in an improbably long drinking straw. During the growing season, a large deciduous tree can suck up as much as 100 gallons (400 liters) of water in a day, just to keep its vital juices flowing.

But if the roots cannot find enough water to satisfy this hectic demand, the slender threads of liquid in the phloem will break, or cavitate—often with an audible pop—and circulation to part or all of the upper tree will be cut off. Death, or dieback, follows. This is exactly what happened, on a large scale, during the drought of the Dirty Thirties when, to cite one well-documented case,

The gossamer-winged butterflies of the family Lycae-nidae are among the most familiar of insects, small, darting glimmers of metallic blue or coppery orange that flutter from flower to flower and spend the summer feeding on nectar. Yet behind all this sweetness and light lies a surprising story of danger, pay-offs, and protection.

Many species of gossamer-wings are reared by ants. Take the Edwards' hairstreak, for example, a tawny little butterfly with a vivid blue-and-orange tail and a faded orange streak on the underside of its hindwing. Edwards' hairstreak is found on the eastern plains, from far-southeastern Saskatchewan to north-central Texas, and eastward across the continent to the Atlantic. So far as is known, it is found only in places that are home both to oak trees and to a particular species of red-and-black ant called *Formica integra*. The oak trees are important as egg-laying sites (in wounds and under rough bark) and as a source of food for the caterpillars. As the flat, slug-like caterpillars creep up the tree trunk to feed on leaf buds, ants scurry along with them and fight off attacks by predators and parasitic insects.

At first, the butterfly larvae have little to offer the ants in return for their services as bodyguards. But as the caterpillars develop, they acquire glands that secrete honeydew, a sugary solution that the ants eat. It's at about this stage that the ants increase their contribution as well, by constructing shelters for their "herd." These are anthill-like cones of dirt and debris—4 to 10 inches (10 to 25 centimeters) in diameter and 2 to 8 inches (5 to 20 centimeters) tall—that appear at the base of caterpillar-infested oaks. The caterpillars, by now nocturnal, feed in the branches at night, with an ant guard at their side, and retreat to the safety of the shelter at first light. As many as 114 caterpillars have been found in a single "byre," together with dozens of attendants.

In midsummer, the caterpillars form pupae, usually in or near the byre, and the first adults emerge in late June or early July. Although some gossamer-winged butterflies are able to reach adulthood either with or without the assistance of ants, Edwards' hairstreaks appear to be entirely dependent on this alliance.

between 50 and 60 percent of all trees in Kansas and Nebraska succumbed to water stress, and most of the survivors were left stunted and partially black-ened. In northern Texas and Oklahoma, where the death toll was even worse, as few as two or three trees in ten survived the decade of drought.

For a tree on the sunbaked prairies, survival is a question of supply and demand. On the supply side, trees reach out for water by producing ambitious systems of roots, which are often almost as massive as the plant's aboveground growth. Contrary to a common misconception, most tree roots are relatively shallow—no more than a few yards, or meters, deep—and they rarely tap into the permanent moisture of the water table. (Even roots can drown if they're held under water and deprived of oxygen for too long.) Instead, the plants

The bur oak, a rugged, slow-growing tree with a deep taproot and thick, fire-resistant bark, is found in tall grasslands and prairie-and-oak savannas. Although individuals may live for a thousand years and attain an enormous girth, the trees seldom exceed a height of 70 to 80 feet (20 to 25 meters).

push out laterally through the rain-watered upper layers of the soil until their rootstocks extend well beyond the circumference of the branches. Green ash, for example—an eastern woodland species that occurs midway across the plains—puts out roots to a distance of some 50 feet (15 meters), but rarely pushes them more than 3 feet (1 meter), deep. The more arid the habitat and the more drought-resistant the tree, the more far-reaching the roots tend to be. Thus, the gnarly little bur oak—one of the most drought-hardy species of trees on the tall-grass prairie and one of the few that carried on merrily during the thirties' drought—typically produces roots that are as deep as the tree is tall and more than twice as extensive. To be specific: a 43-year-old bur oak that had struggled up to a height of 20 feet (6 meters) was found to have pushed its roots out laterally over a distance of 41 feet (12.5 meters).

On the demand side, trees reduce the rate at which water vaporizes and streams away from the surface of their leaves. Deciduous trees, for example, take the radical approach of shedding their leaves in autumn, entering near dormancy, and then bursting forth with new vigor at the time of the spring rains. Conifers have reduced the surface of their leaves to the barest minimum, thereby cutting their water throughput, slowing down their metabolism, and allowing them to photosynthesize sluggishly throughout the year. In addition, the leaves and needles of many species of trees are covered with a waxy, waterproof coating, or cuticle, that helps keep moisture from leaking out, except through the portholes of the stomata. And even these pinprick openings can be regulated to a certain degree—opened when the air is moist and cool or shut when it's dry and hot—to minimize the risk of drying out.

Some species of trees, and even some geographical populations within a given species, are better than others at turning off the tap. For example, trembling aspens that grow on semiarid grasslands have a greater ability to adjust their stomata and reduce their water loss than do aspens from other,

> AMAZING ASPENS

Of all the trees that grow on the prairies, none can bounce back from disaster more quickly than the trembling, or quaking, aspen. ("Stand still for half a hour," one biologist was heard to complain, "and those darned things grow right up your pant legs.") The key to their remarkable vitality lies in their roots, which have a grasslike ability to put up new shoots. The root system is typically shallow but extensive, with lateral runners that can reach 100 feet (30 meters) from the trunk and sinkers that descend from them to a depth of 6 to 10 feet (2 to 3 meters). The entire length of the runners is studded with growth points, as many as thirty in a thumb width, every one of which is capable of producing a new stem and, potentially, a new aspen. Such growth occurs in the first few years after a devastating fire or drought, as saplings shoot up by the thousands and then are gradually thinned out through the survival of the fittest.

Although each stem in the grove eventually develops a more-or-less separate system of roots, there is a very real sense in which a grove of aspen is a single plant. Because all the trees in the stand have risen from the same parent stock, they carry identical DNA—they are in fact clones—and share such characteristics as protein content, sprouting rate, and insect resistance. Each bluff also responds to the seasons at its own genetically programmed pace, so related trees tend to green up and drop their leaves in synchrony. Since aspens are monecious (they produce pollen and seeds on separate male and female plants), all the members of a particular clone are usually also of the same sex. The largest known aspen clone—and the largest organism currently alive—is a stand of 47,000 male stems in the Wasatch Mountains of Utah.

Aspens do produce seeds, but these flights of cottony fluff only rarely take root and produce new plants. The species basically maintains itself without the genetic mixing and innovation that come through sexual reproduction. Instead, each grove sticks with its own time-honored formula, producing shoots that transmit the parent DNA unchanged into the future. Although individual stems seldom survive longer than a couple of centuries, some clones are thought to have sustained themselves for between 1 million and 3 million years.

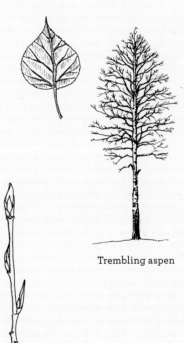

Trembling aspen

Aspen twig
with bud scars

less-challenging environments. But despite these refinements, a typical tree is still a large and thirsty organism that often finds itself tested by the supply-demand equations of the open plains. This helps explain why prairie woodlands are often tucked into folds of the land, close to the water table and protected from sun and wind, where the margins of life, by prairie standards, are most generous.

Where the Trees Are

Until recent times (before agricultural settlement began), the limits of life on the Great Plains were largely set by weather. The invisible contours of the climate—especially large-scale gradients of temperature, precipitation, evaporative demand, and wind—were mapped onto the landscape as visible differences in vegetation. The same west-to-east trend in precipitation that governed the transition from short-grass through mixed-grass to tall-grass prairie had a parallel influence on the height, density, and vigor of woody vegetation. In the dry short-grass ranges of the western plains, for instance, the brush tended to be scrubby, sparse, and tough, dominated by slow-growing, drought-hardy species such as sagebrush and juniper. And in the extreme southwest, where low precipitation was compounded by searing heat, the land was invaded by shrubs from the desert chaparral, including honey mesquite, creosote bush, and the ferociously barbed cat's claw mimosa. Even with this intrusion, as little as 3 percent of the western plains was wooded.

Moving eastward into the mixed-grass zone, the shrubbery gradually became more lush, until the draws were thick with bushes such as choke-cherries, hawthorns, and wild roses, and even the grassy uplands were studded with clumps of snowberry and silverberry (also known, respectively, as wolf-berry and wolf willow). By the same token, the banks and bottomlands of the major rivers, which had been covered with grass farther west, became shrubby and then heavily treed as they cut east through the mixed grasslands. By the time they reached the tall-grass prairie, these streamside, or riparian, woodlands had taken on the character of full-fledged forests, with willows and cottonwoods along the shore, hardwoods such as green ash and American elm (with their understory of shrubs) on the lower slopes, and bur oak at the top of the banks, where they faced off against the bluestems and other grasses. These serpentine strips of riparian woodlands contributed to the estimated 15 percent of the tall-grass region that, under natural conditions, was covered by woody plants.

Green ash

American elm

facing page: A stand of trembling aspen provides a rich tangle of habitats, from the lush understory to the upper branches.

The globular fruits of the wild prairie rose provide nutritious food to a wide range of animals and birds, including deer, elk, sharp-tailed grouse, prairie chickens, rabbits, skunks, and squirrels.

If one left the river valleys and headed still farther east—beyond the basins of the Red and Missouri rivers, toward the eastern hardwood forests—the country went through a passage of ecological uncertainty, in which it was never quite clear if the trees or the grasses were in ascendancy. Still tracking along the same west-to-east increase in precipitation, the tall grasses were increasingly invaded by stands dominated by hickories and oaks until these groves eventually coalesced into a dense deciduous forest. This grassland-to-forest transition occupied a broad arc that swept south from Minnesota to Missouri and then (under the sway of moist weather rolling in from the gulf) swung to the southwest, following the curve of the Ozarks in the direction of Texas. There, at the southern terminus of the transitional zone, lay the Cross Timbers Forest, a picturesque savanna of weather-beaten oaks and junipers interspersed with glades of grass that once sprawled over almost 20 million acres (8 million hectares) in southwestern Kansas, Oklahoma, and central Texas. Although most of the trees were logged off in the early 1900s, significant fragments of the primordial Cross Timbers remain, including gnarled cedars that date back to the sixteenth century.

WOODLANDS AND SAVANNAS OF
THE GREAT PLAINS REGION

| | | | | | Aspen Parklands

Edwards Plateau Savannas

Southern Prairie-and-Oak Transition

Northern Prairie-and-Oak Transition

Black Hills Coniferous Forest

Cross Timbers Forest

0   500 kilometers

A comparable transition zone, or ecotone, is also found on the prairies' northern shore, where mixed grasslands lap up against the boreal forest. Just as the west-to-east increase in precipitation encourages the growth of trees, so the south-to-north decline in average temperatures reduces the evaporative demand and creates conditions that are hospitable to forests. Somewhere around 52 degrees north latitude (about the latitude of Saskatoon), the climate strikes a balance between the requirements of certain midheight grasses—rough fescue, for example, and western porcupine grass—and those of various trees and shrubs, notably, trembling aspens. The result is a pleasantly varied landscape of rolling, grassy meadows and contoured groves, or bluffs, that is known as Aspen Parklands. Extending perhaps 60 miles (100 kilometers) from north to south, the parklands stretch all the way across the Prairie provinces,

> OLD-GROWTH SPECIALISTS

For most of us, the phrase "old-growth forest" conjures up images of towering redwoods and massive firs in the misty gloom of the Pacific rain forest. We're much less likely to think of scraggly old juniper or oak trees, out on the burning plains, that have taken four hundred years to reach the height of a two-storey building. Yet stunted and bent as they are, these venerable prairie warriors deserve not only the respect due to age but also consideration for their increasing rarity.

Take, for example, the twisted trees of the Texas Hill Country. Originally an open savanna of mixed woods interspersed with meadows of grass, the region was heavily logged in the late 1800s. Then, thanks to fire suppression, woody plants pushed back in, often in the form of dense, monotypic, pollen-spewing stands of the native "cedar," or Ashe juniper. Despised by ranchers, developers, and allergy sufferers alike, Ashe juniper was soon listed as Public Enemy No. 1, to be attacked with all possible vengeance.

Yet Ashe juniper has its uses, most notably to a bird called the golden-cheeked warbler. A close relative of the more widespread yellow warbler, this species can be recognized by appearance (dark body, yellow face with black eye stripe), by song ("bzzzz layzee dayzee"), and by location (remarkably,

it nests only in central Texas). Most at home in extensive tracts of mixed woodlands, the golden-cheek feeds mainly on caterpillars and other insects found on deciduous trees. But it also has an absolute and specific requirement for mature Ashe junipers, aged twenty or thirty and up, that have begun to develop shredding bark. The female warbler uses these fibers, bound together with spiders' silk, to build her soft, gray, perfectly camouflaged nest. As shaggy old junipers have become scarce, so have the golden-cheeks, which were granted emergency protection under the Endangered Species Act in 1990. Predation by blue jays from encroaching urban forests (Austin, San Antonio, Waco) and parasitism by cowbirds are compounding the problems, which do not offer easy solutions.

Golden-cheeked warbler

from the foothills of central Alberta to the Canadian shield, and south through Manitoba into northwestern Minnesota. There the broken stands of aspens shade imperceptibly into the savannas of hickory and oak that extend, with remarkably few interruptions, south to the Cross Timbers.

Thus, for the most part, prairie woodlands sort themselves along the gradient of the four cardinal directions, north, south, west, and east, in accordance with their ability to withstand water stress. On a finer scale, they tuck themselves into the prairies wherever the lay of the land either improves the supply of water to the roots or (through shelter and shade) reduces the evaporative demand on the foliage. That is why, for example, the north-facing slope of a coulee is often densely tangled with brush, while the exposed southern slope supports nothing but a carpet of forbs and grasses.

Ice Age Relics

The tensions between supply and demand also help to account for another little-noticed feature of the prairie landscape. You'll be scooting across the flatlands when suddenly the way ahead is obstructed by an abrupt rise. As the road climbs the escarpment, the view tilts toward the light and a phalanx of bristling conifers appears against the sky. Why, you may well wonder, would an evergreen forest sprout on the top of a rocky ridge, smack dab in the middle of the prairies? Yet this phenomenon is repeated in dozens of places across the Great Plains, from the white-spruce forests of the Cypress and Black hills to the ponderosa pines of Pine Ridge to the junipers of the Caprock Escarpment, among many others.

White spruce

These mysterious scarp woodlands are relics from the past. At the end of the last Ice Age, the climate was cool and damp, and a dense coniferous forest stretched across the breadth of the continent. A dark mantle of spruce extended from what are now the Canadian prairies south through the Dakotas into the central states, while pine woods appear to have flourished on the southern plains. But as the chill of the glaciation gradually lifted, the climate eased into a drier and warmer phase, marked by more frequent droughts, and the boreal forest was forced to retreat to the north. Within remarkably short order—a blink of the cosmic eye—the forests of the Great Plains either surrendered directly to grasses or else gave way first to deciduous trees and then to prairie.

By the time the transformation was over, coniferous forest could only be found on the crowns of the tallest breaks and ridges. High enough to

Ponderosa pine

> FOSSIL FORESTS

Relics of the late Ice Age forest are strewn across the Great Plains, rather like flotsam from an ecological shipwreck. In the Sweet Grass Hills of Montana, for example, stands of Rocky Mountain Douglas-fir are found alongside hybrid (white x Engelmann) spruce, growing just where they were left stranded thousands of years ago. The main populations of all three species are now far to the north and west, in the mountains and the boreal forest. Similarly, scientists working in Wyoming and Nebraska have discovered relict stands of hybrid poplars (cottonwood x balsam) in parts of the Niobrara Valley, where the two species had not been in contact since the late Pleistocene.

The distribution of these living fossils has helped researchers to get a sense of the extent of the prehistoric forest. For instance, the fact that there are moose in the Turtle Mountain/ Pembina Hills region of Manitoba and North Dakota has been taken as evidence that the northern plains were once covered by extensive woodlands. By the same token, the flora and fauna of the Black Hills, with their disjunct populations of white spruce, paper birch, moose, black bears, and other boreal species, speak of a time long past when coniferous forest flowed out of the Rocky Mountains and across the Great Plains.

But the late Ice Age is not the only geological episode that has left its mark on the present. There was also an era, roughly 5,000 years ago, when the climate was both moister and warmer than it is now. Under these conditions, southern-adapted species, including a round-eared, silky-haired rodent called the eastern woodrat, were able to migrate north. (More commonly known as a pack rat—from its habit of stuffing shiny objects into its nest—the woodrat looks like a handsome, oversized mouse and is not closely related to the introduced Norway rat.) Although most of the population subsequently retreated southward, a hardy remnant still persists on the wooded banks of the Niobrara River and its tributaries in north-central Nebraska, 100 miles (150 kilometers) from any other members of the species.

catch the rain and snow, and cooler than the grasslands below, these uplands created a microclimate in which the trees could retain a toehold. The thin mineral soils of the ridges—more suitable for conifers than for grasses or other plants—probably also gave the trees an advantage. But the factor that made the biggest different to the scarp forests was their top-o'-the-world location. Historically speaking, the greatest threat to prairie woodlands, apart from prolonged drought, was the fierce heat of grass fires. Where better to find refuge than atop a natural firebreak, an outcropping of safety in a world of flame? The present distribution of scarp woodlands therefore probably represents the limits of wildfires past, a kind of high-tide line beyond which the danger did not pass.

Trees on the Move

For the woody plants that grew down on the grasslands, by contrast, there was no chance of escape. Despite a variety of adaptations for withstanding occasional burns—corky, fire-resistant bark (bur oak); seeds that are stimulated by heat (ponderosa pine); roots that put out new shoots to compensate for fire damage (many poplars and oaks)—few species of trees can survive frequent, intense fires. Grasses, by contrast, are basically born to burn. Not only do they produce a tinder-dry thatch of dead foliage that lights with the slightest spark, but they are equipped to rise from their own ashes. The buds, or meristems, from which they put up new growth are tucked down at the surface of the ground, where they are protected from serious harm. But woody plants, which are inclined to reach for the light, hold their buds on the tips of their branches, where they are exposed to the flames and are sorely vulnerable to fire damage.

In the days of the buffalo prairie, frequent grass fires conspired with severe drought and occasional intense grazing to limit the spread of woody plants. In the sometimes-lengthy interludes between die-backs, trees and shrubs were often able to take advantage of cool, moist weather to extend their reach onto the grassy plains. But no sooner were the trees established than some random bolt of lightning would set the prairies aflame, killing shrubs and most trees except in the humid coulees and river valleys. On the tall-grass prairies, in particular, where catastrophic droughts were relatively infrequent, trees and shrubs might have taken over completely if it hadn't been for the erratic but inevitable return of lightning.

Prairie fires were also set by Native people, who used burning as a tool to hold back the brush and maintain the grasslands as pasture for bison. But with the end of the buffalo ecosystem and the introduction of agriculture, prairie fires were suppressed and trees began to make a slow but steady advance. This expansion of woody vegetation has now been documented in almost every ecoregion across the Great Plains, from the Aspen Parklands of central Alberta (where a 60 percent increase in the area of brush was noted between 1907 and 1966) to the Flint Hills of Kansas (where patches of woodland protected from burning have enlarged by 250 percent since the 1850s) to the dry grasslands of west Texas (where the area covered by mesquite has more than tripled locally in just over a century). At the same time, various species of junipers are intruding on western rangelands, ponderosa pines have expanded from the Niobrara

Eastern red cedar

River valley into the Nebraska Sand Hills, and dense stands of eastern red cedar dominate the once-grassy slopes of Iowa's Loess Hills. These examples could be multiplied many times over.

Extrapolating from local experience to calculate the effects on the Great Plains as a whole has not proven to be simple. Estimated area of presettlement woodlands *minus* acreage lost to clearing and other development *plus* postsettlement expansion of woody growth *equals* the sum of many unknowns. Still, the available evidence strongly suggests that the Great Plains region is now more heavily wooded than it was two hundred years ago. The big, unanswered question is, why? Some researchers have suggested that the changes began back in the days of the cattle barons of the nineteenth century, when the prairie was badly damaged by overgrazing. As the hold of the grass was weakened, patches of bare soil were exposed, creating openings where woody plants could grow. Once the brush was established, it was able to expand. Or perhaps the cattle not only broke the sod but also sowed the seeds by ingesting the fruits of trees and shrubs (notably, honey mesquite) and distributing them in their droppings. Alternatively, it is possible that global, rather than local, forces have been at work. Recent evidence suggests that rising concentrations of carbon dioxide in the atmosphere, due to the burning of fossil fuels, may be encouraging the growth and expansion of trees and shrubs by giving them a physiological advantage over nonwoody plants. But whatever other factors have been involved, the suppression of the natural fire regime has certainly played a central role in encouraging the intrusion of woody plants into the grasslands.

In the Riparian Zone

The transformation of grasslands into woodlands is far from trivial. It represents the difference between ground squirrels and tree squirrels, meadowlarks and robins, sharp-tails and ruffed grouse, between the stripped-down economy of the drylands and the richer options of a more sheltered, and sheltering, wooded environment. Of all the habitats on the Great Plains, none is more biologically productive—or more subject to disturbance—than the sinuous stands of trees and shrubs that line the major creeks and rivers. Although these riparian forests occupy only about 1 percent of the region, they provide living space for somewhere between 60 and 90 percent of the species of vertebrates (reptiles, amphibians, mammals, and birds) that occur on the prairies. For example, researchers working in northeastern Colorado were amazed to discover that

82 percent of all the local birds could be found down by the river. This finding helps to account for the delight of walking along a prairie river on an early morning in spring, as a dozen different bird songs shimmer above the canopy. For anyone who has experienced this pleasure, it will come as no surprise to learn that, in both numbers of species and numbers of individuals, these woodlands are home to some of the richest avian communities in North America.

But the resources of the riparian zone do more than provide habitat for an abundance of wild animals and birds. Over the years, they have also proven highly attractive to humans, from the seminomadic people of the Plains Woodland tradition, who gardened along creeks in Colorado some 2,000 years ago, to the village-dwelling Mandans and Hidatsas, who, as recently as the 1880s, were growing corn, beans, and squash in plots along the Missouri River and its tributaries. With the intrusion of industrial society, the demands on the riparian zone have grown ever-more intense, as we look to the river valleys for farmland, rangeland, roadways, reservoirs, and river-view subdivisions. When these

A pair of yellow warblers tend their nest.
Arthur Savage photo

localized impacts are overlain on the global effects of fire suppression and climate change, the riparian community can be expected to respond in conflicted and complex ways.

Even the most innocent changes can trigger profound effects. For instance, what could look less stressful than a herd of cud-chewing cattle bedded on a riverbank, watching the water slip quietly past? Yet wherever they are present, cattle present a major threat to the health of riparian ecosystems. Unlike the native-born bison, which were well adapted to life on the dry plains, cattle evolved in more temperate environments and are drawn to treed valleys in search of water, shade, and forage. Although moderate grazing, by a small number of cattle for a short period of time, may not cause noticeable harm, heavy grazing inevitably leaves deep scars. Too many hooves in too little space soon pound the place to death, as the animals foul the water, destabilize the banks, and trample or chew the upcoming crop of woody plants. Traditionally written off as sacrifice areas by the cattle industry, riparian woodlands have only recently been appreciated as a rich resource for wild animals and plants. Thanks to this new awareness, range managers are promoting the use of artificial water points, streamside fencing, and other strategies to protect these "hot spots" in the living landscape.

The sparky black-billed magpie is a familiar resident of woodlands across the plains, from the riparian zone to farm shelterbelts and urban plantings. A bright-eyed opportunist that eats both plant and animal foods—including songbird eggs and young—the magpie is despised by people who do not understand the role of predation in the ecosystem.

Unfortunately, damage caused by other means is often more difficult to correct. In the central United States, for example, corridors of hardwood forest once spread across the broad floodplains, or bottomlands, of rivers and creeks, including (among others) the Milk, the Marias, the Wildhorse, and the mighty Missouri. The riparian forests of the Missouri River, for example, opened out in the Dakotas and, like the river itself, grew broader and more majestic as they flowed south, expanding from a width of about half a mile (800 meters) in the north to a span of almost 19 miles (30 kilometers) as they neared the river's mouth. Today, this irresistibly flat and fertile plain has been mostly converted to farms. Between the Nebraska state line and the

Mississippi River, more than 80 percent of the lands that were forested before settlement are now planted to crops like corn and sorghum.

Farther upstream in the Dakotas, the outcome has been similar, though the means have been more complex. Here, too, forests have been lost to clearing: for instance, almost 60 percent of the native woodlands between the Dakotas' Garrison and Oahe dams are now under cultivation. As for the remainder, most of the once-wooded floodplain now lies under the waters of lakes Sakakawea and Oahe and other impoundments. Although the shores of the reservoirs are fringed with patches of brush, these fragmented

Box elder/
Manitoba maple

facing page: With
large, heart-shaped
leaves that whisper
and rattle in the wind,
plains cottonwoods
are among the most
talkative of trees.

woodlands are neither as extensive nor as diverse—nor as rich in species of birds—as the original forests were. And even in places where remnants of the native forest have survived (for instance, along undammed stretches of the river west of Williston, north of Bismarck, and east of Yankton), things are not looking good. The problems first surfaced in the mid-1970s, when an analysis of cores from trees in the Garrison-to-Bismarck stretch of the river, downstream from the Garrison Dam, showed that they were not growing as quickly as they had twenty years earlier, before the dam was constructed. Somehow or other, the impoundment of the river was affecting the health of riparian ecosystems over distances of as much as 60 miles (100 kilometers).

How could a stationary wall of earth and concrete exert this kind of influence? The answer was that it was doing what it had been designed to do: controlling the flow of the river and preventing flooding. By depriving the riparian forests of a well-timed influx of water from spring floods, the dam had cast a shadow over them. In response, all the major species of trees in the forests, including green ash, box elder (or Manitoba maple), and American elm, were suffering from reduced rates of growth. And then there were the cottonwoods.

The dominant trees of the riparian zone across the prairies—and often the *only* species on the arid plains of Alberta, Montana, and other points west— cottonwoods are dependent on the energy and flux of flood for their very survival. Each spring, a typical grove of cottonwoods pumps out billions of cottony, wind-borne seeds, most of which end up drifting forlornly around the country. Only a tiny fraction find the conditions that they need for germination and growth: bare soil, full sun, and plentiful moisture. Where better to meet these requirements than in the wake of a spring flood? As the river recedes from the floodplain, a terrain of newly deposited sandbars and clean-scoured banks is exposed, creating ideal conditions for cottonwood seeds to sprout. (That is why cottonwoods often grow in bands and arcs of same-age trees, each representing a historical moment of opportunity.) But if spring floods are constrained and seed beds are not produced, the trees lose their only chance at reproduction for an entire year.

As Wallace Stegner once pointed out, "western history is a series of lessons in consequences," and the consequences for cottonwoods have, for the most part, not been good. Because their rate of reproduction is reduced, the trees can't compensate for natural mortality, and many populations are collapsing.

(Cottonwoods are old by age thirty, and few last longer than a century.) The surviving stands of cottonwoods along the Missouri, Bighorn, Milk, South Saskatchewan, and several other rivers are all dying off, and there are few ambitious young saplings to take their places. As early as 1981, for instance, just a couple of decades after the Waterton and St. Mary rivers of southeastern Alberta were dammed, the number of cottonwoods along their banks had declined by 23 percent and 48 percent, respectively. By contrast, along the neighboring Belly River, which was not restrained by dams, the cottonwoods continued to reproduce and sustain themselves.

Where the trees are dying off, they are being replaced, in the west, by grassy banks or, farther east, by species of trees that are not dependent on seasonal flooding. These include both native trees such as ash, maple, and elm and invasive, introduced "weeds" such as tamarisk (salt cedar) and Russian olive. But even if other trees do grow up to fill the gap, the absence of cottonwoods is a loss to the roughly one-third of prairie birds that rely on them for food and shelter. Because (like other members of the poplar tribe) they support an exceptionally rich fauna of insects, they are an especially valuable resource for insect eaters such as warblers and woodpeckers. And because of their propensity for heart rot (another poplar attribute), they are an important source of nesting habitat for woodpeckers, chickadees, tree-nesting ducks, and other cavity nesters. Cottonwoods also produce fat, juicy leaf buds that serve as a vital wintertime food for fat, juicy browsing birds such as sharp-tailed and ruffed grouse, which teeter improbably in the upper branches.

The Case of Braided Rivers

Inevitably, on a planet that always has a surprise or two up its sleeve, there is a twist to this story—a special case in which the same suite of human actions has brought about an equal but opposite effect. Along certain rivers in the west-central plains, notably, the Arkansas in eastern Colorado and the Platte system of Wyoming, Colorado, and Nebraska, the curtailment of spring floods has had the unexpected effect of *promoting* the growth of cottonwood forests. To reproduce successfully, cottonwoods need bare, flood-scoured seed beds for germination. In succeeding years, as young saplings, they also need a respite from flooding. A so-called meandering river like the Missouri throws up banks and sandbars as it carves and recarves its course, thereby creating slightly raised beds on which some of the saplings, some of the time, find safe haven. But a

Bufflehead at nest

facing page: Riparian forests and other mixed woods provide habitat for ruffed grouse. By beating his wings rhythmically against his body, this male will thump out a "come-hither" call to attract breeding females.

Arthur Savage photo

The sun sets over the braided—and increasingly brushy—channels of the Platte River in Nebraska.

shallow, braided river like the Platte runs and ripples across its streambed and, under natural conditions, is likely to scour out any seedlings in its channel. Even if a cottonwood seed does manage to put down roots, it will probably be swept away by the river long before it can reach adulthood.

Or at least this is how things used to work, before the dams went in. According to land surveys from the middle to late 1800s, the banks of all three Platte rivers—the north and south forks and the Platte River itself—were originally covered in grass, with only scattered clumps of brush and little or no cottonwood forest. But by the 1950s, the waters of the Platte system had been impounded behind several dozen dams, and peak flows were dramatically reduced. In response to these changes, each of the three rivers dwindled into

progressively narrower channels, thereby abandoning much of their streambeds and exposing a series of beaches, or natural levees, on which trees and bushes could become established. By the time the channels stabilized in the late 1960s, the streams had been reduced to between one-fifth and one-twentieth of their historic widths, and their once-open beds were scrawled with willows and lined with a gallery of towering cottonwoods.

Once set in motion, the wheels of ecological change have just kept on rolling. Because the rivers are now in balance with their flows and are no longer narrowing, they have effectively stopped producing nursery sites for cottonwood trees. It follows that these forests will soon suffer the same fate as the cottonwoods everywhere else: they will die off and be replaced by mixed hardwood forests. This coming transition is already signaled by the advance of Russian olive throughout the Platte drainage, a silvery foreshadowing of future change. But despite these continuing adjustments, the forests that have sprung up along the Platte system appear to be here to stay, at least for as long as the rivers are constrained.

Plains cottonwood

Go West, Young Bird

From a bird lover's point of view, it may be hard to understand why anyone would object to the expansion of woodlands along the Platte or anywhere else on the Great Plains. The more trees have invaded the grasslands over the last hundred-odd years, the more species of birds have moved in, either as summer residents or as visiting migrants. No matter how the expansion of woody growth was triggered—whether through overgrazing or climate change or the suppression of fire and flood—the result has been a diversification of the local environment. The addition of woody vegetation brings a complex vertical dimension to the living world that extends from the ground up, through the shrub layer, into the sheltered midcanopy and the treetops. At every level,

> RIVER OF CRANES

Congruence: The state of being in agreement, harmony or correspondence, from the Latin "con," or with, and "grues," or crane.

Every spring, between late February and early April, about 375,000 sandhill cranes—roughly 80 percent of all the sandhills in the world—descend on the Platte and North Platte rivers in central Nebraska. There, at the halfway point of their annual journey from the Llano Estacado of New Mexico and Texas to their nesting grounds in Canada, Alaska, and Siberia, they settle in for several weeks to rest and refuel. By daylight, the birds disperse to forage in wet meadows and cornfields along the river, where they put on weight for the breeding season to come. At night, they return to the river to roost among its shallow, braided channels. Packed in shoulder to shoulder, they rest at half-alert, stirring the darkness with the tremolo of their high-pitched, gurgling voices.

With a fossil record from Nebraska that dates back more than 6 million years, sandhill cranes are the oldest of the world's living birds. The Platte River, by contrast, is a mere youngster that traces its origins to the end of the most recent glaciation. From this it follows that the congruence between the cranes and the river arose sometime in the last 10,000 years. Recently,

however, this harmony has been threatened by the construction of more than 200 dams and diversions on the Platte system. With peak spring flows now well below what they were even during the 1930s drought, the river is no longer able to maintain the flood-fed wetlands and the open, unvegetated sandbars that the cranes require. Forced out of historic roost sites by the encroachment of woody plants, the cranes are now heavily reliant on a few key stretches of the river system. Despite a conflict-ridden effort to "mitigate" the impacts of water management, current research suggest that cranes are showing signs of physiological stress. In particular, the birds' ability to store energy for their onward journey (their daily rate of fat deposition) has declined by as much as 50 percent since the late 1970s. Meanwhile, the total count of spring migrants has dropped by 125,000 birds, or 25 percent, since the fat years of the early 1980s. The consequences for the endangered whooping crane and piping plover, which also rely on the Platte River, have not been thoroughly investigated.

Sandhill cranes

trees and shrubs provide habitats to a particular community, or guild, of breeding birds that is adapted to specific conditions of light, humidity, and shelter. In addition, woodlands also offer a range of options along the radius of each wooded stand, from the shadowy stillness of the interior to the bright and breezy margins. For an ovenbird, a secluded forest interior. For a catbird, a shrubby edge. For an oriole, the airy heights of the upper branches.

Life responds to opportunity, and the expansion of woody vegetation across the plains has opened up new opportunities for dozens of species of birds that were seldom, if ever, seen here by early settlers. If the first generations of prairie naturalists could peer down from heaven, binoculars in hand, they would be astonished by the flood of recent arrivals and range extensions. On the northern plains, for example, the western kingbird (a flashy gray-and-black flycatcher with a lemon-yellow front) was originally restricted mainly to wooded river valleys in the drainage of the Missouri and the South Saskatchewan rivers. But since the late 1800s, the birds have ventured out of the valleys and, discovering that the world has changed, have expanded through North Dakota and across the plains and parklands of the Canadian prairies. Like land birds crossing an ocean, they have hopped from island to island of wooded habitat, laying claim to a country that in the past had denied them access.

The change that made all the difference for western kingbirds was not so much the expansion of natural islands of brush as the sudden appearance of artificial forests. Ever since the first Arbor Day was celebrated in Nebraska, in 1872, the people of the Great Plains have eagerly bent to the task of cultivating what one prairie arbori-enthusiast referred to as "missionaries of culture and refinement." By which he meant woody plants. Over the decades, the settlers and their descendants have planted literally hundreds of millions of trees and shrubs, including so many shelterbelts that if they were placed end to end, they would reach around the equator and partway back again. Between 1935 and 1958, for instance, a combined total of 32,000 miles (52,000 kilometers) of these rural windbreaks were planted on the grasslands of Canada and the U.S., providing a literal hedge against soil erosion. Meanwhile, in both rural and urban settings, the little house on the prairie had become the little house in the woods, as tree-loving people altered the landscape to create habitat for themselves. As these plantings matured, they also provided habitat for western kingbirds and other species of birds, which used them as stepping-stones in their progress across the grasslands.

When Euro-Americans began to think about settling permanently on the bald-headed prairies, many people found the idea bewildering. How, they wondered, could a civilization that had risen in the dappled woodlands of Europe and eastern North America put down roots in a land that was unable to support a proper forest? The lack of trees on the prairies was widely seen as a mark of deficiency: no lumber, no fuel, no rain. No nothing.

From the 1890s onward, countering this aspersion became the personal mission of a pioneering Nebraska botanist (the Johnny Appleseed of the plains) by the name of Charles E. Bessey. As Bessey saw it, the Great Plains grasslands represented the ruins of a prehistoric forest that had been brought low by bison and grass fires. If only the trees could be restored, he thought, the climate would improve—precipitation would increase—and life on the plains would be easy.

With this utopian vision in mind, he and his colleagues set out to plant a mixed-wood (largely pine) forest on a tract of windswept sand and grass near Halsey, Nebraska. Although Bessey died in 1915, his eccentric dream lived on, and in the 1930s his efforts were extended by the tree-planting brigades of the Civilian Conservation Corps. Today, the Bessey Division of the Nebraska National Forest sprawls over 31 sections (80 square kilometers) of land and provides habitat for a wealth of woodland species that would not otherwise be present in the Sand Hills, including cardinals and other songbirds, wild turkeys, and white-tailed deer.

Unfortunately for Bessey's project, however, his namesake forest cannot sustain itself and has to be augmented through planting and other human interventions. In the end, it seems that the grasslands were not a failed forest after all but a natural and fitting response to a midcontinental climate.

The list of woodland birds that have extended their breeding range on the prairies since settlement began includes such once rare and now at least locally common species as tree swallows, mourning doves, downy woodpeckers, house wrens, chipping sparrows, black-billed magpies, American crows, American robins, merlins, Mississippi kites, and red-tailed hawks, among many others. The combined effect of these changes is suggested by a recent study of the birds in northeastern Colorado, which showed that almost 90 percent of the species found in the area today did not occur there at the turn of the twentieth century. The vast majority of the newcomers are highly adaptable, go-anywhere woodland species with broad distributions in other parts of the continent and no special affinities to the prairie region.

facing page:
In North Dakota, the population of the wild turkey has risen from zero in 1950 to an estimated 20,000 in 2010. Following an initial introduction program, the birds have established themselves along the Missouri River and in the Pembina Hills, among other places.

Sexual Tensions

Although most of us would not want to do without the robins bobbing on our lawns, the nuthatches hanging from our feeders, or the chipping sparrows chittering in the shrubbery, the abundance of woodland birds on the prairies—and the ecological changes that they represent—are also a subtle menace. What appears at first glance to have been an unequivocal enrichment of life turns out, on closer examination, to pose a covert threat to the specificity, the very prairieness, of the prairie region. For instance, in the old days, before woodland birds could leapfrog across the grasslands at will, the open plains (especially the sparsely treed western rangelands) acted as a barrier between eastern and western species. On one side of the great grassy divide, in the ecotone along the eastern hardwood forests, for example, one would find species such as blue jays, eastern bluebirds, and indigo buntings. On the other side, toward the mountain forests of the west, were closely related birds that occupied much the same ecological niche as their eastern counterparts yet were obviously different: Steller's jays (with black crests rather than blue), mountain bluebirds (with blue breasts rather than red), and lazuli buntings (with rich splotches of cinnamon adorning their plumage). Presumably first separated thousands of years ago, at the time when the Ice Age forests were forced off the Great Plains and the reign of grasses began, these populations had been held apart long enough for each to develop its own genetic specializations. Thus, merely by preventing regular contact, the grasslands had driven the process of evolutionary change and sparked the creation of new species.

No more. These days, with riparian woodlands extending into the prairie heartland along the Platte River, and woodlands dotted hither and yon, eastern and western species are meeting head on. Of a total of thirteen east-west pairs of woodland birds that converge on the Great Plains, seven have been studied in depth and all have been found to interbreed freely along their zone of contact. In some cases, the interchange appears to be stable, suggesting that the two gene pools have been in intermittent contact over the centuries and now coexist peacefully. For example, the yellow-shafted (eastern) and red-shafted (western) phases of the northern flicker meet and merge in the Missouri borderlands of Montana and North Dakota, where they produce orange-shafted offspring just like the ones that John James Audubon saw in the same region in 1843. By contrast, Baltimore and Bullock's orioles have been jostling back and forth across the prairies for the last twenty-odd years, with Baltimores extending 120 miles (200 kilometers) west across the Canadian prairies and up the Platte River, and Bullock's orioles pushing strongly eastward across southern Kansas. Along this fluid zone of interaction, the two species interbreed, producing birds with mix-and-match markings and raising the possibility that the two

> IN THE HYBRID ZONE

This chart (adapted from a study by Paul Ehrlich and his associates at Stanford University) lists thirteen pairs of closely related species of birds that were isolated on opposite sides of the continent by the emergence of the Great Plains grasslands some 10,000 years ago. Long held apart by un-bridgeable expanses of prairie, these species have recently been brought back in touch by the expansion of deciduous woodlands. Although some of these long-lost cousins may have become sufficiently different from each other that they cannot interbreed, others are still similar enough to hybridize freely.

Facial markings of the red-shafted flicker, *left*; yellow-shafted-flicker, *center*; and their hybrid, *right*

Eastern Form	Western Form	Hybrids	Studied?
eastern screech-owl	western screech-owl	rare	no
yellow-shafted flicker	red-shafted flicker	common	yes
red-bellied woodpecker	golden-fronted woodpecker	rare	no
great crested flycatcher	ash-throated flycatcher	unknown	no
eastern wood-pewee	western wood-pewee	unknown	no
Carolina chickadee	black-capped chickadee	rare	no
tufted titmouse	black-crested titmouse	common	yes
eastern bluebird	mountain bluebird	very rare	no
rose-breasted grosbeak	black-headed grosbeak	locally common	yes
indigo bunting	lazuli bunting	common	yes
eastern towhee	spotted towhee	common	yes
eastern meadowlark	western meadowlark	rare	yes
Baltimore oriole	Bullock's oriole	common	yes

distinctive populations might eventually merge into one. If this were to happen, the diversity of the world's bird life—the range of answers that life can offer to the questions posed by change—would be incrementally diminished.

Meanwhile, another group of prairie birds is already vanishing, and the influx of forest species has likely not been helping. The grassland specialists—ground-nesting birds like meadowlarks and dickcissels, pipits, bobolinks and

The bright, melodious song of a male western meadowlark announces his presence to the world. Though still common and widespread, both western and eastern meadowlarks have suffered declines during recent decades.

short-eared owls—have suffered dramatic population losses during the past forty or fifty years, and probably for longer. (The available data only go back to 1966, when the annual Breeding Bird Survey, or BBS, was initiated.) During that period, the "grassland bird indicator," a composite measure that reflects the populations of two dozen prairie-loving species, has declined by almost 40 percent. In part, this trend is the result of loss of habitat to agriculture and urbanization. But the birds are also declining even where significant fragments of native grasslands still exist, and the encroachment of woody plants is a likely culprit.

In a recent study, researchers working in the cattle country of northwestern Oklahoma compared the advancing tide of woody vegetation (as revealed by a thirty-year sequence of aerial photographs) with the downward trend in

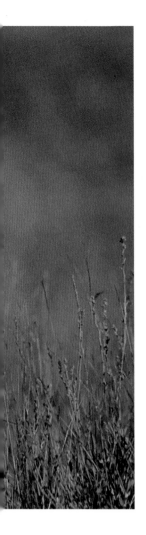

grassland birds (as revealed by thirty years of BBS data). They detected a strong, inverse relationship between the two variables. The more trees and shrubs, the fewer grassland birds. But does this relationship hinge on simple appropriation of space, or are other, more devious interactions also at play? Could it be, for example, that the spread of trees has tipped the ecological balance against grassland birds and in favor of their enemies or competitors?

In their search for answers, researchers have pointed the finger of suspicion at one of the prairie's own native-born residents. The brown-headed cowbird, or "bison-bird," as it might better be called, is a small-to-medium-sized member of the blackbird tribe. (Females are drab; males brown headed and splendidly iridescent.) In the olden days, cowbirds lived with, and even on, the bison, following them wherever they wandered and feeding on the insects that flew up from their hooves. Although food was seldom wanting, this nomadic lifestyle did impose a considerable cost: the birds had no chance to settle down, build nests, and raise young. Instead— remarkably—they acquired the ability to drop their eggs in the nests of other songbirds and fly off, leaving their offspring for somebody else to incubate, feed, and bring up. Accordingly, cowbirds are classified as obligate brood parasites, because they are entirely dependent on other species to perform nursery duties.

The obvious losers in this transaction are the unwitting hosts, which often put so much effort into feeding their greedy guests that they cannot meet the demands of their own young. In one instance, only 12 percent of meadowlark chicks were found to have fledged in parasitized nests, compared with 26 percent in broods with no cowbirds present. Yet up to a certain point, this interference does not pose any real threat, since the hosts have had thousands of years to adapt to nest parasitism. (Even 12 percent of a suitably large clutch may be all that a species needs to keep its numbers up.) But if the rate of parasitism increases, the balance will shift and the host species will slowly but surely begin to slip.

This appears to be exactly what happens, in some instances, when woody plants expand into the grasslands. Each intrusion increases the length of the "shoreline," or edge, along which grass meets woods, the very zone in which opportunists like the cowbird are most successful. The nearer a nest lies to a stand of trees, the greater the chance that a cowbird will find it and come calling. This may be either because cowbirds concentrate their search along grassland-woodland edges (where they have the best of two habitats) or because they perch on exposed branches, overlooking the prairies, as they search for places to drop their eggs. In either case, improved success for cowbirds will likely come at the expense of already beleaguered grassland specialists. Thus, the apparently thriving populations of dickcissels on small, brush-invaded prairies in southwestern Missouri are suffering from poor reproductive success, probably because of nest parasitism.

If You Plant It, They Will Come

Cowbirds are not the only danger that lurks in the shrubby interface between grasslands and woods. These productive ecotones also provide habitat for a variety of mostly small, nosy, nest-raiding mammals, including skunks, weasels, raccoons, foxes, opossums, and Franklin's ground squirrels. Again, the greater the penetration by woody plants, the greater the threat to ground-nesting birds. (Increased predation by small mammals appears to be a major factor in the decline of some grassland specialists.) For the edge-loving mammals, by contrast, the altered face of the prairies, with its patchwork of grass and brush, offers an irresistible wealth of resources. The whole place seems to shout, "Come and get it!" The list of mammalian predators that have responded to this call and expanded their range includes raccoons and opossums (which have been heading north), eastern spotted skunks and gray foxes (slowly pushing west), cougars (heading east along major watercourses), and red foxes (striking out wherever they can in every direction).

It is not only predators that have benefited from the increased interspersion of wooded habitats. A whole range of other mammals—from tree-roosting red bats and tree-climbing white-footed mice to tree-gnawing porcupines—have also taken advantage of the shifting tree line. Fox squirrels, a bright-eyed, bushy-tailed tribe of arboreal rodents that was once restricted to open woods along the easternmost fringe of the plains, have pushed steadily north and west in recent years, scampering from river valley to farmyard to city street. By 1979,

facing page: The fox squirrel subsists on a diet of acorns, supplemented with green shoots, flowers, insects, and even the occasional frog. Acorns and nuts are buried as food for the winter, and any that are left grow up to renew the forest.

for example, they had established themselves in Regina, by way of the Souris River. Since then, they have expanded east up Thunder Creek and north across the Qu'Appelle River, to within calling distance of the boreal forest. Meanwhile, their cousins in Nebraska have been pushing west along the South Platte River to occupy the shelterbelts and riparian forests of eastern Colorado. If you plant a tree, you end up with a squirrel.

But perhaps the most remarkable story of all is that of white-tailed deer. Now a familiar sight almost everywhere on the prairies, these alert, graceful creatures were originally native to the hardwood forests of eastern Canada and the U.S., where they made themselves at home in grassy meadows and open woods. As recently as the mid-1800s, this was the white-tails' world and they seldom, if ever, ventured westward onto the Great Plains. Although a small population seems to have colonized the riparian forest of the Missouri River— prime habitat, one might think—they were rare even there and were seldom noticed by early exploration parties. Instead, the journals of prairie exploration are filled with references to the abundant herds of elk that, in those halcyon days, flourished in the riparian forests and hill country. (Although the prairie

The mule deer, above, can be recognized by the black markings on its tail and by its big, muley ears. The white-tailed deer, shown on the facing page, takes its name from its flaglike, waggling tail, which it raises in alarm.

subspecies of elk was decimated in the 1880s, due to hunting and habitat loss, Rocky Mountain elk have since been introduced to the Black and Cypress hills and several other refuges. The native subspecies persists in parts of southern and central Manitoba and Saskatchewan.)

It was only when the explorers struck out across the grasslands that they really started to notice deer: a small herd of does at the mouth of a brush-filled gully, a lone buck silhouetted along the crest of a rocky ridge. But even here, these sightings were not of white-tails. These animals were mule deer, a typically western species that ranged—and still ranges—from the woody draws and breaks of the prairies through the Rockies and all the way to the Pacific. Most at home in rough, open country marked with tangles of scrubby brush, the mule deer is characterized by its large, alert ears and hair-trigger nerves. It is perpetually poised to bound, or "stott," with its peculiar, stiff-legged gait, up and over obstacles and out of harm's way.

The white-tailed deer, by contrast, is adapted to a gentler landscape of meadows and leafy groves, where a frightened deer can melt into the shadows. At the first hint of danger, the flaglike tail shoots up, sending a conspicuous, waggling warning to friends and relatives. And away they all run, into the nearest woody cover. (Female white-tail and mule deer both typically live in same-sex groups, often with their mother, sisters, and aunts. Bucks tend to be solitary, except during the rut, and seldom associate with close relatives.) Although of little benefit on open grasslands, this defensive strategy has proven remarkably well suited to the mosaic of croplands, hay fields, and scattered groves that has sprung up on the prairies since settlement. Despite a turn-of-the-century crisis, in which both mule and white-tailed deer were almost exterminated by overhunting, white-tails have come back in abundance and extended their range across the Great Plains and beyond. They are now by far the most numerous wild ungulate in North America, with a continental population that likely stands at four or five times what it was when the *Mayflower* landed.

Meanwhile, the deer of the west—the mule deer—have been placed at a disadvantage by the changing landscape, including the expansion of invasive trees, and have suffered widespread population declines since the 1950s. In Montana, for example, where mule deer traditionally outnumbered white-tails by a ratio of 2:1, the species are now about equally common. Since muleys are typically able to dominate white-tails (to put them on the run), this shift presumably is not the result of competition for food or other resources. Instead, the declining fortunes of the mule deer appear to be symptoms of the loss of the shrubby, open country that the species loves.

eight

THE NATURE OF FARMING

The most tragic conflict in the history of conservation
is that between the conservationists and the farmers ...
It is tragic because it is unnecessary.

WENDELL BERRY, "HOPE," 2002

THE FIRST HINT that something strange was about to happen
was the sound: a loud metallic hissing, almost a clattering, that
seemed to rise out of the very ground. We were out on the bald-
headed prairie, scuffing along a dirt track, which, so far, had led
us through a heat-stressed field of hay. In front of us lay a farm
dugout, a pond that had been gouged crudely out of the earth,
its banks a moonscape of heaped dirt. The soil around the crater
was gray and thin, bare of vegetation and brittle with drought.
With every step, tiles of earth crumbled beneath our boots.

The ominous, high-pitched buzz assaulted us from every
side, spinning up from our ankles and ricocheting off the dirt
banks. Was this a warning? Should we turn around and head
back across the field, to the safety of our vehicle?

It couldn't be rattlesnakes, not unless there were dozens of
them, circling around us. But that was too farfetched to be credited. What else? A vague thought stirred in my mind. Hadn't I
read something somewhere about "rattlesnake grasshoppers"?
Could mere insects really set the air ringing with such menace?
Apparently so, for a moment of observation confirmed that
the margins of the dugout were jumping with grasshoppers—
some with yellow bodies or scarlet wings but most of them so
drab that they blended into the gray earth. I could stare straight
at them and not even guess they were there. Then, the second

Carlinian snapper

facing page: The spur-throated grasshoppers, or melanoplines, are the most numerous and diverse group of hoppers on the Great Plains. Members of this group (which includes the lesser migratory and two-striped grasshoppers, among many others) can be recognized by the spiny bump between the front legs.

I moved a muscle, these armored forms broke cover, flung themselves into the air, and spurted away with a loud clacking of their wings, only to fall back to the ground at some little distance and disappear again. As they landed, they sent another cloud of noisemakers racketing into flight, and so on and on, until the whole place was jangling. These were the creatures that had set our nerves on edge and made us consider turning back.

But before there was time to relax, a sudden burst of noise pierced through the ambient buzz, an intense and aggressive-sounding rattle. I looked up to see what I assumed must be a grasshopper suspended in midair above the top of the bank, its wings blurred with effort. The insect hung there, clattering loudly, ablaze with reflected light, for what seemed to be several long minutes, before dropping back to the ground and out of sight. A few minutes later, another frenetic performer leapt into the sky and enacted the same remarkable feat. Alarm gave way to amazement. What was going on here? What kind of grasshoppers were these and what on Earth were they doing?

If you have ever tried to catch a grasshopper on a sun-warmed slope, you will appreciate the absurd scene that followed. Time and again, clumsy human hunter stalked spring-loaded insect—run up close, crouch, and pounce—only to come away empty-handed. But finally, an unwary specimen, preoccupied with munching on a dry stalk of grass, permitted itself to be captured and carried away for examination. A brief check in a guidebook confirmed that this was, in fact, a rattlesnake grasshopper or, as it is perhaps more correctly known, a Carlinian snapper. The insect in hand was a female, recognizable not only by the pronged egg-laying organ, or ovipositor, at the end of the abdomen but also by her quiet, businesslike demeanor in the field. You won't find a female Carlinian snapper making a spectacle of herself. That risky behavior is left primarily to the territorial males, which (as near as anyone can tell) perform their showy display flights to attract potential mates. In so doing, they demonstrate not only their stamina but also their daring, since by drawing attention to themselves, individuals also run the risk of being noticed by predators. Tellingly, a small flock of songbirds lined the far bank of the dugout, and the ground was strewn with sleek gray-and-white flight feathers, presumably shed by one or another species of grasshopper-hunting songbirds.

It is worth emphasizing that these life-and-death dramas were being played out not on native prairie but in an environment that, over the preceding hundred-odd years, had been intensively cultivated. This field was a worst-case

A Swainson's hawk, one of many species that has made itself at home in farming country, sets off to hunt for rodents in a hay field.

situation; land that was totally unsuited for farming—stony, dry, and thin—had been plowed up, eventually abandoned as cropland, and only later planted to crested wheatgrass and reclaimed as a marginally productive hay field. The banks of the dugout, by contrast, had not been reclaimed at all; they were as barren as they had been on the day they were scooped out. Yet in the midst of this human disturbance and disappointment, the wild vigor of the grasslands, as seen in the ancient lineage of the Carlinian snapper grasshoppers, had found room to express itself, in all its strangeness and surprise and fascination.

The North American Agricultural Revolution

Cropland is a wild place that has been partially tamed. By definition, it has been stripped of its native vegetation and seeded to one or another of the small, select group of plants that humans have brought under cultivation. (In the entire world, only about 70 species of plants are commonly grown as crops; by comparison, there are 5,000 wild plants on the Great Plains alone.) Agriculture was first introduced to the prairies at least two thousand years ago by immigrants from the hardwood forests of what is now the eastern United States. Sometime around the beginning of the current era, circa AD 1, these people began tending small plots of corn and sometimes beans, along the tree-lined river valleys of the tall-grass region, from western Missouri to Nebraska. (At one point, around AD 1000, they had extended their influence west to Colorado and north to the Red River lowlands of Manitoba.) By crafting an economy that depended not only on small-scale agriculture but on hunting and gathering as well, the people of this Plains Woodland tradition sustained themselves on the grasslands for a thousand years.

Sometime between AD 900 and AD 1400, however, disaster apparently struck (perhaps in the form of a severe regional drought) and the Plains Woodland culture appears to have died out. Yet even as this group was declining, another wave of farming-and-hunting people was already on the move. Known as the Plains Gardeners or Plains Village culture, these folks established themselves in small, scattered clusters of earth lodges along rivers and creeks from the upper Missouri River south to the Republican River and its tributaries, in present-day Nebraska and Kansas. There they lived by hunting, fishing, gathering, and small-scale agriculture, growing not only beans and corn, as the Plains Woodland people had, but also squash and pumpkins. Known from archaeological strata that date back to AD 1000, this tradition was maintained by peoples such as the Pawnees, Mandans, and Hidatsas for more than eight hundred years. It was only in the mid-1800s, after the introduction of smallpox, that the villages and gardens were abandoned.

Obviously, this pioneering agriculture was responsible for a small, localized loss of natural habitat. But as minuscule islands of human activity in a vast expanse of native grassland, the gardens did not pose a threat to the functioning of the ecosystem. The natural disturbance regime of the prairies was still operating at full force. Wildfires swept freely across the landscape, as random as lightning and wind, removing dried-out vegetation and making way for fresh

Spring wheat belt
Winter wheat belt
Corn belt

0   500 kilometers

MAJOR CROPS OF THE GREAT PLAINS

green growth to push in. Bison, too, tracked across the plains and generally stirred things up, leaving behind a free-form mosaic of grazed and ungrazed plants, dusty wallows, and splotches of dung. By constantly creating new habitats, these natural forces kept the prairie in a continuous state of renewal, able to the meet the vital requirements of its many and varied inhabitants.

The introduction of intensive agriculture, however, brought these natural processes to a halt. Within less than a human lifetime, between the 1870s and the 1920s, wildfire was brought under control by the settlers, the bison were killed off, and a mind-boggling expanse of grassland was brought under

WHERE HAS ALL THE GRASSLAND GONE?

	Historic (ha)	Current (ha)	Decline (%)	Protected (%)
TALL GRASS				
Manitoba	600,000	300	99.9	*
Iowa	12,500,000	12,140	99.9	.01
Kansas	6,900,000	1,200,000	82.6	*
Minnesota	7,300,000	30,350	99.6	1.0
Missouri	5,700,000	30,350	99.5	1.0
Nebraska	6,100,000	123,000	98.0	1.0
North Dakota	1,200,000	1,200	99.9	*
Oklahoma	5,200,000	*	*	*
South Dakota	3,000,000	449,000	85.0	*
Texas	7,200,000	720,000	90.0	*
MIXED GRASS				
Alberta	8,700,000	3,400,000	61.0	.01
Manitoba	600,000	300	99.9	.01
Saskatchewan	13,400,000	2,500,000	81.3	.01
Nebraska	7,700,000	1,900,000	77.1	*
North Dakota	13,900,000	3,900,000	71.9	*
Oklahoma	2,500,000	*	*	*
South Dakota	1,600,000	*	*	*
Texas	14,100,000	9,800,000	30.0	*
SHORT GRASS				
Saskatchewan	5,900,000	840,000	85.8	*
Colorado	*	*	*	*
South Dakota	179,000	*	*	*
Oklahoma	1,300,000	*	*	*
Texas	7,800,000	1,600,000	80.0	*
Wyoming	3,000,000	2,400,000	20.0	*

Of all the rodents, none has made the transition to farmland more successfully than the pretty little deer mouse. Largely nocturnal, it takes shelter in underground tunnels and subsists on a diet of seeds and large numbers of insects and insect larvae, mites, and ticks.

cultivation. The most hospitable and fertile lands—those with the deepest soils, most generous rainfall, and gentlest slopes—naturally bore the brunt of this agricultural revolution. By the time the dust settled, virtually all of the tall-grass prairie, with its rich tangle of grasses and flowers, had disappeared, replaced primarily by an ever-expanding monoculture of corn. (Originally native to Central America, corn is a warm-season grass, like the native species it displaced, and is suited to the growing conditions in the sunny and well-watered eastern regions of the Great Plains.) In the mixed-grass zone, by contrast, the dominant crop is wheat, a cool-season grass that was first domesticated in the Middle East. Together with oilseeds and a number of specialty crops, such as lentils and chickpeas, wheat now dominates between 15 and 99.9 percent of the mixed grasslands, the exact percentage depending on local growing conditions. Only land that is utterly unsuited for crops has been left in native grass, a circumstance that is increasingly common in the south and far west. But even there, on the drought-prone ranges of the short grasslands, about 30 percent of the landscape is under tillage. (These estimates are drawn from *Terrestrial Ecoregions of North America*, a 1999 publication of the World Wildlife Fund.)

THERE'S MORE TO A SKUNK THAN ITS STINK

Much of the current research on farming and wildlife focuses either on insects, especially those that are troublesome to farmers, or on birds, especially those that are troubled by agriculture. But there are many other creatures that live in farming country and that are equally interesting.

Take striped skunks, as an example. These pretty, placid little carnivores are found everywhere across the Great Plains, especially in farming country. Once restricted to wooded streambeds or shrubby meadows, they are now entirely at home in a landscape of grain fields, fencerows, and shelterbelts. Active primarily at night, they pad down farm lanes and windrows, stalking grasshoppers, digging up wasp nests, or sniffing at lumps of excrement to nose out beetles. Insects are their staple foods, but the skunks are opportunists and will readily consume fruits and vegetables, as well as mice, frogs, lizards, carrion, and the eggs and young of ducks and other ground-nesting birds.

Although often described as solitary, striped skunks are actually quite companionable. Breeding females spend the spring and summer tending families of twittering young, usually four to six in a litter. Their nursery is a sheltered den under a building, inside a hollow log, or dug into the ground, sometimes the abandoned burrow of a badger, coyote, or fox. Completely helpless at birth, the kittens cannot walk until they are five weeks old, but they grow up fast and follow their mother on hunting trips by the age of two months, one plumy tail after another. A few weeks later, the youngsters are ready to fend for themselves, and the family disperses.

As winter closes in, however, the skunks den up again, this time in groups that include one male for every roughly half dozen females. Sometimes as many as twenty animals curl up together to wait out the cold, not quite hibernating but very drowsy and slow. This peaceful atmosphere is disturbed briefly in late winter, when first the older females and then the yearlings come into estrus, but calm is quickly restored and the animals slumber on until the weather improves. In due course, the mated females choose their birthing dens and the cycle is repeated.

The transformation of these broad expanses of wild prairie into farmland has been an abrupt and radical shift. Where the cycle of life was once driven by natural processes, much now depends on the activities of humans. The seasonal regimen of field work—the annual round of tillage, seeding, spraying, and harvesting—creates a new rhythm of disturbance that native species must cope with as best they can. At the same time, the substitution of a handful of cultivated plants for the riot of native grasses, forbs, and shrubs has reduced both the range of food and the diversity of habitats available to wildlife, with predictable consequences. It is no great surprise that cultivated fields support fewer species of native prairie animals and plants than do native grasslands. For instance, according to one study, a small (27-acre, or 11-hectare) remnant of unbroken land in Saskatchewan provided habitat for twice as many species of butterflies as were found on the entire surrounding acreage of cropland.

These young barn swallows have almost outgrown their nest under the rafters of an abandoned grain bin.
Arthur Savage photo

Similarly, research done in Nebraska found thirteen species of birds nesting on untilled land but just two in a nearby cornfield. In other words, 85 percent of the native grassland bird life was missing.

But for species that can adapt to the changed conditions, these human-dominated "agroecosystems" are a new and wide-open frontier. In addition to the crops themselves (and what are they but birdseed, after all?), farming has introduced a number of wildlife-friendly features to the landscape of the Great Plains. Shelterbelts now provide habitat for tree-dwelling species such as squirrels and woodpeckers; fences and telephone wires crisscross the country with a network of elevated perches. Skunks den under abandoned buildings and porcupines winter in rusted-out cars, finding shelter in places where it did not exist before. These examples could be multiplied many times over. Although farmland is not native prairie and does not support the full array of species that were here before, it very clearly has a vitality all its own. Simply by going about their business, prairie farmers have provided living space for wild animals and plants. In return, they enjoy a daily access to the natural world that makes them the envy of many city dwellers.

Unwanted Guests

The presence of wildlife in an agricultural ecosystem is not always welcome. Farming often appears to be a struggle against nature, just one darn losing battle after another. Take the fight against weeds, for example. Weeds are wild plants that seed and sustain themselves mainly, and sometimes entirely, on cultivated land. Because they are able to fit their life cycle into the routines of agricultural work, they can compete with crops for sunlight, water, nutrients, and space and greatly reduce yields. Although a few weedy species, like quack-grass, are perennials, which sprout up from fragments of root, most are annuals, which rely on seeds to reproduce. They tend to be fast growing—able to mature between farm operations (for example, before planting or after the harvest is off)—and to produce large crops of big, well-nourished seeds. Fecundity is their main defense against frequent disturbance. If every plant produces scores of viable offspring, at least a few of them are likely to avoid being weeded out and survive to produce their own superabundant generation of seedlings.

Perhaps because the original prairie grasslands were largely composed of perennials, with few annual plants, it is uncommon for native species to have weedy characteristics. Instead, our common weeds are mostly exotic species that were accidentally introduced during the transition to agriculture. Though the bane of the farmers' existence, their nutrient-rich seeds now provide a reliable, year-round food resource for small mammals and birds, from mourning doves and horned larks in the summer to snow buntings in the winter. Weeds also provide nourishment for a wide variety of insects, especially larval forms. For example, the leaves of Canada thistle (a plant that, despite its name, was introduced into Manitoba from Europe in the 1880s and has since spread across the plains) have become a staple food of painted lady caterpillars, so much so that the adult forms are now sometimes known as thistle butterflies.

Insects that consume weeds may be welcome; not so those that eat crops. Controlling herbivorous insects is the other major challenge facing agriculture. Again, most species that become pests have been introduced from abroad, including such "bad guys" as the Hessian fly, the European corn borer, the English grain aphid, and—a relative newcomer—the Russian wheat aphid. Many

native insects, by contrast, have no chance of becoming economic threats, because they are adapted to feed on species or families of plants that are of no commercial importance. To cite one typical example, the female Harris' checkerspot butterfly lays her eggs exclusively on a wild plant called the flat-topped aster, the only food her caterpillars will accept. Nothing else (not even one of the other seventy-odd species of wild asters) will meet their requirements. So unless flat-topped asters someday become a marketable crop, this is one insect—among many—that will never trouble farmers. Similarly, the larvae of the great spangled fritillary and a number of related butterflies feed only on violet leaves and are unlikely to do anything more worrying than nibble on pansies.

Painted lady caterpillar

Nevertheless, many native insects have made the leap from natural prairie to cultivated fields. Among them is a small black-and-yellow-striped wasp (about the size of a large mosquito) that was originally associated with awned wheatgrass, Canada wild rye, and other hollow-stemmed native grasses. The female wasp bored a hole into these stems and laid her eggs inside. After the larvae hatched, they fed within the stem until autumn, when the plant began to dry out and the insects headed toward the ground. A short distance above the base of the plant, they gnawed a notch around the stem, causing it to bend or break and form a stub. Here, beneath a plug of plant matter, the larvae spun a silk-lined nest to protect themselves through a winter of diapause, or insect hibernation. In spring, they resumed development and matured into flying adults, which emerged from the now-dead stems to lay their eggs in the new year's growth of grasses.

Harris' checkerspot
butterfly

facing page: A butterfly and a swarm of beetles have all been attracted to feed on the head of a thistle.

When wheat was introduced to the Great Plains, the wasps apparently did not take to it right away or at least did not immediately cause noticeable damage. In Manitoba, for example, they continued to rely primarily on their natural hosts until 1906, a decade after intensive wheat farming had begun. That year, however, perhaps because of a failure of the native grasses, flights of stem-boring wasps began to lay their eggs in wheat. Because they are inclined to lay eggs in the species of plant from which they hatched, they quickly formed a lasting attachment to fields of wheat. Their success was enhanced by the absence of several of their natural control agents—predatory insects and parasitic wasps—that for some reason did not make the transition from native grasslands to crops.

And so a pest was born. The wheat-stem sawfly, as this insect somewhat misleadingly came to be known, went on to cause massive crop losses on the northern plains in the 1950s and 1960s. Although its importance diminished after solid-stemmed varieties of wheat were developed, it has recently staged a remarkable comeback. In the past, sawflies infested only spring-sown wheat, a crop that grows through the summer and is harvested in the fall, giving the insects time to emerge, lay eggs, hatch, and initiate larval development during the growing season. But they could not survive in fall-sown, or winter, wheat, which matures more rapidly and dries down before the sawflies are ready. Lately, however, evolution appears to have pressed fast-forward and produced a new strain (or perhaps species) of sawfly that emerges a month earlier than before, allowing it to colonize winter wheat as well. And so the process of adaptation continues.

Hopper Heaven

Of all the insects that have benefited from farming, none have been more successful—or more devastating to human interests—than certain species of grasshoppers. During severe infestations, they may consume crops, gardens, shrubbery, and everything else in their path until, as the book of Exodus puts it, "not any green thing" is left. People who lived through the Dirty Thirties may still remember walking through a green slime of swarming insects. No wonder we tend to think of them as intruders. Yet grasshoppers are not intruders; they are native herbivorous insects that, over the last 50 million years or so, have evolved in concert with the natural vegetation of the grasslands.

Although plants may look like passive victims, they are really small green warriors armed to fight off attack. Grasses, for instance, are girded with strands of abrasive silica that make them hard to chew and digest; leaves coated with hairs have much the same effect. Many forbs, including sagebrush and other aromatics, are laced with substances that block digestion or prevent nutrients from being absorbed. To survive, grasshoppers (like other herbivores) have had to come up with ways of overcoming these defenses, whether by developing more powerful mandibles and larger gizzards for grinding up tough plants or by producing antidotes to specific plant toxins. No one species can rise to every challenge.

The result has been the evolution of hundreds of species of grasshoppers— more than four hundred in western Canada and the United States alone—each of which has overcome one or more sets of obstacles and gained the ability to feed on one or more types of vegetation. Some grasshoppers are so specialized

Wheat-stem sawfly

that they can eat only a single type of native grass, while others are restricted to
one family of plants. The rattling Carlinian snapper, for example, is thought to
feed mainly on wheatgrasses, and Turnbull's grasshopper (the farmer's friend)
eats only Russian thistle and other related plants. These highly specialized hoppers seldom, if ever, become pests, mainly because they are such fussy eaters.

By contrast, the small number of grasshopper species that do infest croplands tend to be generalists, with wide habitat tolerances and broad food preferences. For example, the clear-winged grasshopper—a major pest of cereal crops
across the north-central plains—is naturally adapted to eat many types of wild
prairie grasses, including fescues, bromes, and bluegrasses. And what are wheat
and barley but highly nutritious kinds of grass? The introduction of agriculture
has provided clear-wings with mile upon endless miles of top-quality sustenance, far better than anything available to them under natural conditions. The
network of mowed road margins has also benefited them by providing ideal
sites for egg laying. (When laying, a female grasshopper bores a hole into the
soil with her abdomen, releases a number of cream-colored-to-orange eggs, and
then covers them with a protective secretion. This pod is the overwintering
stage for all the pest species.)

Turnbull's grasshopper

Most grasshoppers breed in late summer and die before winter sets in, leaving the next generation to overwinter as eggs in the soil. There are, however, a number of notable exceptions, among them a smallish gray-to-green insect with bold geometric markings that is somewhat confusingly known as the brown-spotted range grasshopper. This species lays its eggs in midsummer, whereupon they either (a) sit there for an entire year and hatch out the following July—the usual pattern on the northern plains, though no one knows why—or (b) get busy and hatch out immediately, as one might expect and as commonly happens in the central and southern states. In either case, winter soon closes in on the young hatchlings and they are forced to take shelter under leaf litter or are buried alive in the earth.

Come early spring—usually sometime in April—the half-grown hoppers emerge from their winter hideaways, resume feeding (on a diet of sedges and grass), and proceed to grow and develop. But no sooner do they become active, than they are set upon by a rush of breeding birds that spill across the plains for the nesting season. So many mouths to feed, so many bite-sized grasshoppers to stuff into them. Because these early-season grasshoppers are available in May and June, when other insects are scarce, they serve as a critical resource for many species of grassland birds. By the time the summer is over, a single pair of prairie songbirds and their nestlings can consume up to 11 pounds (5 kilograms) of insects, including as many as 149,000 grass-hoppers.

Brown-spotted
range grasshopper

Another notorious pest species, the two-striped grasshopper, has had an even better run. A real go-anywhere, eat-anything herbivore, it is equipped to feed not only on grasses but also on a wide range of broad-leafed plants, including everything from mustard to sowthistle and from ragweed to leadplant. It can even eat timber milkvetch, a wild legume (found in the mountains and the Black Hills) so poisonous that a few mouthfuls can kill a cow. This hopper is one tough hombre, a kind of six-legged supervillain. Not surprisingly, it consumes all types of crops, including grains, alfalfa, corn, and garden produce, and lays its eggs in a variety of agricultural environments: roadside ditches, weedy summerfallow, stubble fields, and tame pastures. For species like the two-striped grasshopper, farmland is heaven.

Having provisioned these insects so abundantly, farmers find it no easy task to prevent them from taking over. Grasshopper damage can be reduced by tactics such as early seeding, crop rotation, tillage, and the use of trap strips (green strips planted adjacent to a germinating crop, where grasshoppers can be lured and poisoned). When worst comes to worst, however, producers often end up relying on lethal force, by spraying infested regions with insect poisons. The current generation of insecticides is effective, but it is also crude, killing not only pests but any other insects or spiders that happen to be in the way. And while modern insecticides do not persist in the ecosystem as older control agents (like DDT) did, they can still leave behind a wreckage of collateral damage. One popular preparation, carbofuran, proved to be so toxic to birds—including several dozen California gulls that died after feeding on poisoned hoppers in Saskatchewan—that it was taken off the market in Canada and the United States as a grasshopper-control agent, though it is still used in other countries and for other purposes. And even when these poisons do what they are supposed to do and kill only insects, they still weaken the web of life by wiping out the food supply for insect eaters. Thus, the battle against grasshoppers may unintentionally take a toll on snakes, lizards, mice, skunks, bats, quail, small hawks, and songbirds—all of which need insects to feed their young—among many others.

Ultimately, the course of a grasshopper outbreak is beyond human control. Like other insects, grasshoppers are creatures of the weather. As cold-blooded organisms, they regulate their body temperature by seeking sun or shade (for example, by basking at the top of a stem or hiding under a leaf) to keep their tissues in the optimal functioning range. Warm, dry summers give grasshoppers an advantage, because they don't have to spend so much time soaking up the sun and can put their energy into feeding and growing. If sunny weather persists into late summer and autumn, when most species lay their eggs, the scene is set for an explosive population increase. Under optimal conditions, a single female can produce as many as 250 eggs. If the following spring brings more fine weather, the new generation of nymphs, or hatchlings, will prosper and reproduce in their turn, until the whole place is hopping with hoppers.

About the only thing that can rein in this runaway population is a good spell of rain. By preventing the insects from maintaining their body temperature, damp conditions hinder their development and growth. At the same time, humidity encourages the spread of deadly grasshopper diseases. Of special note is a fungal pathogen called *Entomophaga grylli*, which, when conditions are right, can decimate a population of grasshoppers almost overnight. A strange malady if ever there was one, *E. grylli* causes its victims to clamber into the top of plants just before they die. (Perhaps the illness makes them feel chilly and so they climb up to bask in the sun.) There they cling, so many small, faded cadavers grasping onto stems in what can only be described as a death grip. Forced into the service of their nemesis, the grasshoppers apparently serve as aerial capsules in which the fungus can mature and from which it can broadcast its spores. Although this disease can bring a quick end to a grasshopper outbreak, it requires ten days of warm, wet weather to develop full epidemic force and therefore is not reliable enough to be deployed as a biocontrol agent.

Benefiting the "Beneficials"

Because they are so responsive to the weather, grasshoppers and many other insects have a built-in tendency to experience cycles of boom and bust. Yet in a state of nature, it must have been relatively uncommon for even the most voracious among them to cut loose and reach plague proportions. The resources they required for survival—including their preferred food plants and suitable egg-laying sites—were often limited and dispersed. They had to struggle to meet their needs or, frequently, die trying. At the same time, they were faced with perils that included not only disease but a large and varied assault force of spiders and predatory and parasitic insects. Although these natural control agents have been pushed to the sidelines by high-tech agriculture, they are still out there in the ecosystem.

Grasshoppers, for example, are beset by insect foes, including both beeflies and blister beetles. Although beeflies really do look like bees, all hairy and striped, they only have one set of wings, a characteristic that identifies them as members of the order Diptera, the flies. (True bees, by contrast, always have two sets, fore and hind.) Blister beetles are typical beetles, though with long, narrow bodies that are often clad in wonderfully iridescent or striped wing casings, or elytra. (And, yes, they do contain chemicals that can cause blistering.) Although totally unrelated to each other, beeflies and blister beetles have adopted similar reproductive strategies. They lay their eggs in the soil close to grasshopper egg-pods, a resource that the gravid females likely locate by smell. When the larvae hatch, they wriggle around in the dirt and attack the egg-pods, consuming grasshopper embryos by the dozen. Although not numerous enough to control a severe outbreak on their own, they do their part in limiting grasshopper populations.

The same is true of certain species of tachinid flies, which look like houseflies festooned with bristles or hairs. All members of this group are parasitic,

Beefly

facing page: The banded argiope spider lives in open grasslands. Its wispy-looking webs are sturdy enough to capture and hold grasshoppers, which are thought to be the principal food of this species.

which is to say that they typically lay their eggs on—or sometimes in—the larvae of other insects. There the eggs develop and pass through various larval stages before bursting out of their unwitting host as adult tachinids. Some of the species that infest grasshoppers are live bearers, hatching their eggs in their own bodies and then laying their larvae on, or in, the early nymphal stages of their victims. But however the attack is launched, these parasitic invasions are fatal.

The trick to maintaining populations of these natural pest-control agents is to provide food for them. Even predators and parasitoids have to eat. After they emerge as free-flying adults, beeflies, tachinids, and blister beetles all feed on pollen or nectar from their own preferred flowering plants. In other words, a weed-free, wall-to-wall monoculture cannot support them. Instead, they require, at a minimum, strips and patches of land that are left to grow weedy and wild, whether along roadsides, grassed watercourses, or fence lines. A diversity of plants supports a diversity of "beneficial" insects. For instance, the introduction of wild sunflowers has been known to reduce the population of greenfly on a nearby sorghum field, probably by providing nectar for a particular species of greenfly-killing parasitic wasp. And sometimes the linkages are even more circuitous. There is a certain tachinid fly, for example, that parasitizes the European corn borer, a pest that (as its name suggests) tunnels into cornstalks and destroys them. The tachinid is most effective in the presence of giant ragweed, a plant that is often infested by yet another type of stalk-boring insect. Since the tachinid parasitizes this insect as well, it can survive in its alternate host when European corn borers are rare, and so it is ready and waiting to mount an attack when the pest reappears.

Tachinid fly

It's easy to see how providing food and alternate hosts for beneficial insects would help control pests. But there may well be more to the story than that. When plants are attacked by herbivorous insects, the leaves release volatile compounds into the air, which waft out from their tissues like distress signals. Predatory and parasitic insects can pick up these chemical messages (or smells) and interpret them, perhaps even to the extent of identifying the species of plant-eating insect that is present. If the predator or parasitoid catches the scent of an edible insect or a prospective host, it will home in on the signal and attack, thereby protecting the plant from extensive damage. The more species of plants that are present, the more varied and stimulating their chemical output will be, as if the SOS call were beaming out on every available

wavelength. Certain parasitoids have proven to be more effective in controlling pest infestations in mixed plots of corn, soybeans, and weeds, with their inter-mixed signals of distress, than they are in pure stands of corn, which all send out the same message. Diversity multiplies into complexity in natural systems.

There are more than a thousand species of sweat bees in North America, all of which provide essential services as pollinators.

A Fondness for Beetles and Bees

Of all the predatory insects that live in croplands, none are more successful—or more important in controlling pests—than the legions of beetles. To date, more than 350,000 species of these armor-winged insects have been identified in various parts of the world, making them the largest and most diverse order of living animals. (Asked what he had learned about the Creator from a lifetime spent studying His works, the Scottish biologist J.B.S. Haldane is said to have quipped, "He seems to have an inordinate fondness for beetles.") By and large, the species found in farmland belong to the family Carabidae, or ground beetles,

a group that includes some 40,000 species worldwide. Of these, an astonishing 2,500 species are known to occur in Canada and the United States: that's 2,500 subtly different ground beetles. The most familiar of their number, the sidewalk beetle, is a European species that was accidentally imported by early settlers and that, ever since, has been fast-footing it through lawns and gardens across the continent. Most of the native carabids are variations on the same theme: small, black, shiny, and intensely busy.

Although the majority of carabid species cannot cope with frequent disturbance and are not found on cultivated plots, the number of species that have made the transition to farmland is impressive. For instance, by the simple expedient of setting wide-mouthed jars into the soil and fitting them with funnels instead of lids, a researcher in South Dakota was able to trap 127 kinds of ground beetles on farmers' fields. And while many of these species were rarities, several were exceedingly abundant, with peak, late-summer populations of 4,000 to 6,000 beetles per acre, or 10,000 to 15,000 per hectare. On one extraordinary occasion (when a heavy rain a few days earlier had moistened the soil and created what must have been ideal conditions), one species of tiny carabid topped the charts at an estimated density of 5.6 million individuals per acre (13.8 million per hectare). No matter how you put it, that's a lot of beetles. Yet for the most part, we don't even know they're out there because they hide from the sun by day (in the soil or under litter and leaves) and only come out to hunt in the cool, moist air of evening.

Sidewalk beetle

But under cover of darkness, what carnage! Although a few species of ground beetles feed on plants, most are voracious predators that prowl through the dark, searching for insect eggs, larvae, pupae, and soft-bodied adult forms. While some carabids are thought to be specialists, with their sensory radar tuned to specific prey, others appear to be opportunists that eagerly accept any insect food that comes within reach of their mandibles. As a result, they consume untold quantities of insects that might otherwise damage crops, including wireworms, cutworms, caterpillars, maggots, and aphids, among many others. It would not be overstating the case to say that farmers could not attain profitable yields without the unpaid—and generally unheeded—exertions of ground beetles.

The best way to safeguard the diversity and abundance of these essential "farm animals" is to conserve their habitats. Practically speaking, this means maintaining weedy field margins, grassy watercourses, irrigation ditches,

facing page: In their adult form, Nuttall's blister beetles are as showy as the flowers on which they feed.

roadsides, fence lines, shelterbelts, pastures, wetlands, and other semiwild refuges where the land is subjected to minimal disturbance. Such areas not only provide safe havens during insecticide application—beetles have been seen running toward field edges to flee from toxic sprays—but also serve as sheltered daytime hideaways. (Dense tangles of vegetation offer humidity and shade.) Perhaps most important, these undisturbed patches also provide essential wintering habitat, where beetles can burrow down into the ground and wait for spring's return. The overburden of vegetation provides protection against extreme cold and may also trap an insulating layer of snow, thereby further improving the insects' chances of survival. In one experiment, a raised, grassy berm, or "beetle bank," was found to shelter 100 overwintering carabids in every square foot (1,100 per square meter).

Retaining strips and patches of undisturbed land is the minimal precondition for sustaining the web of life in intensively farmed regions. Although it is true that these small, untended corners can harbor weeds and insect pests, any problems are generally far outweighed by the benefits. Given that the vast majority of native species are either beneficial or neutral to farming, reserving space for them becomes a simple matter of economic and ecological good judgment. Besides pest control, there is the additional plus of free pollination services. Although crops like wheat and corn are pollinated gratis by the wind, many others—including sunflowers, canola/rapeseed, lentils, alfalfa, and peas—must be pollinated by insects, most notably, by bees. Unlike the colonial honey bee (another import from Europe), most of the thousands of species of native bees are solitary nesters. Each female lays her eggs either in holes in wood or, more often, in cells in undisturbed ground, provisioning each with a dollop of a pollen-based mixture. In producing this "bee bread," the female trails pollen from one flower to the next, thereby ensuring that the plants set seed and produce a marketable yield. Thus, simply by conserving areas of semiwild habitat where bees and other wildlife can live, farm people can expect to profit in every sense of the word.

Farming and Birds

If the introduction of intensive agriculture constituted a revolution in the ecology of the Great Plains, prairie farming itself has since undergone radical change. A lot has happened since the horse-and-buggy days. The original patchwork of small, diversified homesteads, with a family on every farm, has

> **FLIGHT OF THE BUMBLEBEES**

Bumblebees are the teddy bears of the insect world, fat, fuzzy, cute-as-bugs, and surprisingly docile. Although they are equipped with stingers, they seem reluctant to use them unless their nests are seriously threatened. Of all the native species of North American bees, the fifty-odd species of bumblebees are the only ones that are social rather than solitary.

Instead of operating as single mothers that provision young on their own, bumblebee queens are assisted by a busy, buzzy work force of helpers. The story begins in the spring when the queen (the only member of the colony to overwinter) emerges from her tiny underground cell and sets out to find a suitable site for her nest. Once she has found what she is looking for—typically a clump of fine vegetation on the ground, in a building, or down a rodent hole—she fashions the nesting material into a ball, within which she deposits both a little wax pot of honey and a lump of pollen-and-honey mixture. The queen feeds from the honey pot and lays her eggs (all female) on the pollen lump, before covering them with wax and incubating them. After about five days, the larvae hatch, feed on the pollen, and begin the month-long transition into adult workers. By the time the first batch is ready for action, the queen has laid a second brood, which the workers then help to rear. And so it goes, for the rest of the summer. Toward the end of the season, when the colony reaches its peak, the queen produces a brood of male and female larvae that, royally provisioned by the workers, grow up to be breeding males and virgin queens. After this sexual generation leaves the nest to mate, the entire colony dies off, except for the young queens, which tuck themselves underground to wait for spring.

Sometimes, however, this simple story line takes a strange twist, and the reproductive bees that fly out from the nest are aliens. In addition to social bumblebees, there are also parasitic forms that require workers to tend their larvae but that cannot produce their own helpers. Instead, a parasitic bumblebee queen enters an active colony, ousts the reigning monarch, and coopts the worker bees, which henceforth help to tend the larvae and pupae of the invading species.

gradually been transformed into a sparsely populated landscape of large-scale, specialized operations. Where once there were fields of oats for the draft horses and pastures for the family cow, there are now entire sections devoted exclusively to the production of cash crops. Bigger machinery has called for bigger fields, with clean straightaways and right-angle turns; it's hard to steer something as big as a bus around the land's natural contours. Squeezed between high costs for fuel and other necessities and ruinously low prices for what they grow, farmers increasingly rely on government income-support programs. Because their entitlements are typically based on acreage under cultivation, they have been pushed to work every square inch of land and to view natural areas as "idle" and "unproductive."

No one knows, in any detail, how these successive changes in the agricultural landscape have affected populations of beneficial insects on prairie farms (though a sharp decline in many species of native pollinators is cause for alarm). Fortunately, however, researchers have been more successful in tracking the fortunes of farmland birds. The greater prairie chicken, for example, was originally a bird of the tall-grass prairie and the neighboring prairie-and-oak savannas. As agriculture spread, this handsome brown-and-buff-barred grouse expanded its range widely, advancing west to the foothills of the Rockies, south to the Gulf of Mexico, and north to the Aspen Parklands. When agriculture was in its frontier phase, this whole broad sweep of country catered to the chickens' needs, combining grasslands for cover with grain fields for foraging. Soon there were more prairie chickens on the Great Plains than there had ever been before. For example, the *Omaha Republican* reported that, on September 6, 1865, two hunting parties led by Captains Kennedy and Hoagland bagged a total of 708 prairie chickens (together with 6 quail, 10 snipe, 14 hawks, 16 ducks, and 2 rabbits). Such one-day totals were not uncommon.

Although protective laws were quickly invoked, enforcement was lax and the population of prairie chickens inevitably began to pay the price. Yet even after the slaughter ended, sometime around World War 1, prairie-chicken numbers did not rebound. By then, the look of the country had shifted from a semi-natural landscape, with small patches of cultivation—perfect prairie chicken habitat—to an intensively farmed ecosystem with small remnants of grassland. The result has been the slow and painful disappearance of the species from four Canadian provinces and eight American states, along with the fragmentation and contraction of the birds' remaining range. Several key populations continue to diminish.

The point here is not just the plight of the greater prairie chicken, as compelling as that is. It is the ongoing narrative of cause and effect, as farming evolves from one stage to the next. Every modification of agricultural practices produces ecological consequences, destroying habitat for some species and, with luck, creating it for others. Thus, the very changes that spelled doom for prairie chickens were to the advantage of other birds, which flocked in to take advantage of novel habitats, including shelterbelts, road allowances, and field margins. Today, several decades further into the story, these strip habitats in their turn have begun to disappear, as farmers strive to push production to the limits. Since the 1930s, it is estimated that between 30 and 80 percent of the field margins in the central United States have been brought under the plow, and the same trend can be seen across the Great Plains in general.

Horned lark

Kildeer on nest

The consequences of these changes are not difficult to foresee. A study conducted in the farm country of north-central Iowa found that the region was home to a fabulous diversity of breeding birds, with close to one hundred species in total. Of these, almost half nested in woody fencerows and windbreaks, a quarter in wetlands and sloughs, and another half dozen or so in hay fields. Removing any of these elements from the landscape removes its resident birds. By the time only the crops and weedy strip habitats are left, only eighteen species remain. And without the weedy margins, there are just three survivors: the vesper sparrow, the horned lark, and the killdeer. (Because these species are adapted to nest on bare or sparsely vegetated prairie and to breed early in the season, often before field work begins, they can succeed right out in the crops where few other breeding birds venture.)

The more the agricultural landscape is simplified, the fewer species of birds it can

support. It's not exactly rocket science. Among the common farmland birds that are now declining in parts of the Great Plains are such stalwarts as American crows, European starlings, house sparrows, mourning doves, common grackles, and American goldfinches.

Abandoned farmsteads not only provide daytime roosts for great horned owls, they also provide shelter for rodents, foxes, and bats, among many others.

At the same time, happily, some recent trends in agriculture appear to be more positive for our feathered friends. One is the widespread adoption of no-till and minimum-till farming methods. Rather than cultivate the land—leaving the soil bare and exposed—the gospel of no-till asserts that the ground must always be covered by vegetation. To this end, stubble and other crop residues are left in place after harvest for direct seeding the following spring, and herbicide spraying, instead of tillage, is used to control weeds. Designed primarily to prevent soil erosion, this technique has also turned out to provide nesting cover for certain species of ground-nesting birds, including meadowlarks,

Until the early 1940s, lesser snow geese made a leisurely passage up the Missouri Valley in spring from wintering grounds on the Gulf Coast to nesting sites in the Arctic, pausing to feed in shallows and marshes along their route. On the return trip in fall, however, most of the population made the journey more-or-less nonstop and were seldom seen along the river. All that began to change in the 1950s, when flurries of snow geese began showing up along the Missouri in both spring and fall, in ever-increasing numbers.

This change appears to have been precipitated not by natural causes but by a shift in farming techniques. Until the 1950s, most of the corn grown on the Great Plains had been harvested by hand, a method that was laborious but thorough. The newfangled mechanical pickers and combines, by contrast, were quick-and-dirty machines that left large areas of the countryside strewn with top-quality goose feed. Thus, the table was set for snow geese (and other waterfowl), which shifted out of the wetlands to gorge on the spilled corn. Thanks to this superabundance, more snow geese survived and went on to produce large broods of equally well-nourished youngsters.

As irrigation allowed corn production to expand, the geese expanded, too, eventually shifting their principal spring migration westward toward central Nebraska. Each spring, not just flurries but blizzards of snow geese spill into the wetlands of the Rainwater Basin and Platte River Valley, where they build up stores of fat and protein before continuing on their journey. Recent research suggests, however, that the geese are not faring as well as they used to, perhaps because of a shortage of food; the efficiency of the corn harvest has been greatly improved. Concern has also been voiced about the shift from corn to soybeans—poor provender for geese—and the introduction of genetically modified varieties. Since genetically modified soybeans are immune to herbicides, they can be sprayed for millet and other weeds, thereby removing an important alternate source of food not only for migrating geese but also for other seed-eating birds and rodents.

pheasants, and grasshopper sparrows. (To return to Iowa again as an example, no-till cornfields in that state have been found to provide nesting sites for a dozen species of birds—up from the expected three—and the nests are packed in at many times the usual density. In this respect, no-till achieves the same benefits as organic farming.)

There is always the worry, however, that these lively no-till fields may ultimately function as ecological traps, first attracting birds in large numbers and then subjecting them to seeding, spraying, and other nest-wrecking disturbances. Just because a species is present doesn't mean that it is managing to reproduce. What's more, if the new soil-holding techniques enable producers to cultivate steep slopes and rough country that could not be farmed in the past, then no-till may be responsible for a net loss of quality habitat.

Fortunately, these ambiguities do not cloud the promise of another tactic: the removal of marginal farmlands from crop production. This process began in earnest during the Dirty Thirties, when many of the dustiest lands in the Dust Bowl were brought under public management as National Grasslands in the United States and as PFRA (Prairie Farm Rehabilitation Administration) community pastures in Canada. Thereafter, progress was sporadic, with periods of reversal, a continuing loss of native grassland, and a gradual extension of the acreage under cultivation. In 1985, however, the cause of biodiversity received an unintended boost when the United States introduced the Conservation Reserve Program (CRP). By contracting with farmers to put marginal farmlands into permanent cover (usually either introduced or native grasses) for a specified term, government officials hoped to protect the soil from erosion and simultaneously put some cash into the pockets of farmers who had been hit by the collapse of world markets. As good luck would have it, the program also ended up offering a helping hand to another group of hard-pressed prairie residents. Grassland birds that for decades had been losing ground—fading away year by year—found refuge in the 30-plus million acres (over 12 million hectares) that were protected from cultivation. Alas, the government giveth and the government taketh away, and these gains are not secure.

nine

LONG-RANGE FORECAST

For a long time it has come back to us in wishes,
this great prairie and these beautiful canyons.

KIOWA ELDER AND U.S. ARMY SERGEANT I-SEE-O, AMARILLO, TEXAS, 1924

IN A CENTURY when the natural world is slowly dying all around us—when wildness has been pushed to the margins—the wide open spaces of the Great Plains are a landscape of hope. Here is an ecosystem that has experienced the full onslaught of modernization in one brief historical instant and that, though battered and torn, still inspires us with its splendor. This is a country filled with light. It is a place where city streets flow out onto the prairie and draw us along until, almost before we know it, we find ourselves rolling down a dusty gravel road, with warm gusts of meadowlark song blowing in through the open window. It is a land where the seasons surge over us like tides, from the sudden upwelling of spring to the languid heat of summer and from the rushing retreat of autumn to the great sparkling silence of winter.

Look up into the darkness of a prairie night and you will see the universe streaming with stars. Suddenly, it becomes possible to picture yourself on the third planet out from the sun, traveling through the mystery and wonder of whatever is out there.

The prairie opens us to the immensities of space and time. Like few other places on Earth, it reminds us that life operates within broad horizons, with sight lines that extend from the past through the present and into the future. Just as the buffalo

> THE SKY IS ROUND

Lakota elder Nicholas Black Elk, Hehaka Sapa, was born on the Little Powder River near what is now the Montana/Wyoming border in 1863. He died on the Pine Ridge Indian Reservation in South Dakota in 1950. This quotation is taken from the book *Black Elk Speaks*, which he published in 1932 with coauthor John G. Neihardt:

> *Everything the Power of the World does is done in a circle. The sky is round, and I have heard that the earth is round like a ball, and so are all the stars. The wind, in its greatest power, whirls. Birds make their nests in circles, for theirs is the same religion as ours. The sun comes forth and goes down again in a circle. The moon does the same, and both are round. Even the seasons form a great circle in their changing, and always come back again to where they were. The life of a man is a circle from childhood to childhood, and so it is in everything where power moves.*

prairie is gone, though not forgotten, the countryside that we see before us is even now being transformed into the living landscape of tomorrow. As we look at the world that we have inherited from our ancestors, it is impossible not to think of the generations who will come after us. The wild prairie that we leave to them will be our legacy.

Admittedly, the trends of recent decades have not been encouraging. Although the big plow-down of the settlement era is behind us, native prairie is still being lost year by year and bit by bit, whether to cropland, wind farms, strip malls or rural subdivisions. According to a recent assessment, about 425,000 square miles (1.1 million square kilometers) of natural grassland have been destroyed in the western United States in the last 150 years. Of these losses, almost 10 percent—42,000 square miles (110,000 square kilometers), an area nearly half the size of Wyoming—were incurred between 1950 and 1990. Although current statistics are not readily available, the destruction has clearly

not stopped. We can rip up 10,000-year-old grassland in an instant, but it is beyond our powers to create it.

Almost as worrying as the outright disappearance of native prairie is the degradation of what is left, whether through overgrazing, fragmentation by energy development, or the intrusion of invasive plants. Not surprisingly, this incremental damage—death by a thousand cuts—is taking a toll on species that rely on wild grasslands for their survival. Take birds, for example. A study released in 2003 by the North American Commission for Environmental Cooperation (CEC) identified thirty-two species of birds that are "highly dependent" on the Great Plains of Canada and the United States, in the sense that more than half of the population is found in the region at some time of year. This group includes not only permanent, year-round residents, such as prairie chickens, but also fair-weather friends, like Sprague's pipits and marbled godwits, which are present only during the breeding season. In addition, there is a

The raucous call of the yellow-headed blackbird is fading from parts of the prairies because of regional declines in the population of this species. Loss of wetland habitat, intensified by drought, may be the root of the problem. Overall, the species is secure.

Once restricted to the foothills and aspen parklands, mountain bluebirds expanded eastward across the plains following settlement. Their initial success was dampened by competition from house sparrows and European starlings (both introduced from Europe). Fortunately, the provision of nesting boxes has helped to ease the situation in many localities, and these bluer-than-sky-blue creatures still add a shock of color to spring on the western prairies.

Arthur Savage photo

short list of species—sandhill cranes and common mergansers among them—that crowd onto the southern plains during the winter. By analyzing data from the Breeding Bird Survey (a standardized count conducted each year since 1966 in both Canada and the United States), the CEC determined that more than 60 percent of the species that rely on the Great Plains are declining in abundance. In contrast, 23 percent of woodland species and about 28 percent of all bird species in North America are experiencing losses.

A similar study in 2005 focused exclusively on upland prairie birds—Swainson's hawks, killdeers, meadowlarks, bobolinks, long-billed curlews, and the like. Of the thirty-seven species under consideration, all but five were found to be in decline. What do these "prairie canaries" have to tell us about the state of our grassland ecosystems? And if prairie birds are suffering, how can we be confident that the other critters out there (the rodents and creepy-crawlies, say, that inspire less interest) are managing any better?

The Western Advantage

By now, you may be wondering what happened to the "landscape of hope" that you were promised a few minutes ago. We're lovers in a dangerous time, and preserving the splendor of the living world is a global challenge. As Bruce

> GRASSLAND-DEPENDENT BIRDS

According to a report entitled the "Importance of North America's Grasslands to Birds," there are about three dozen avian species that are critically dependent on the Great Plains grasslands of Canada and the United States for survival. In each case, at least half of the continental population is found in the region during all or part of the year, often in association with native grasslands and natural wetlands. The list is arranged in declining order of dependency, with birds that rely on the plains exclusively, or almost exclusively, at the top and those that are more widely distributed at the bottom. Asterisks indicate species that have been highlighted by the United States Fish and Wildlife Service as "species of conservation concern."

Year-Round

* lesser prairie chicken
* greater prairie chicken
  sharp-tailed grouse
* McCown's longspur

Nonbreeding season

Smith's longspur
sandhill crane
common merganser

Breeding season

* Sprague's pipit
* marbled godwit
* upland sandpiper
* mountain plover
  blue-winged teal
* Swainson's hawk
* Mississippi kite
  western meadowlark
* Baird's sparrow
* chestnut-collared longspur
* lark bunting
* grasshopper sparrow
* piping plover
  yellow-headed blackbird
  western kingbird
* ferruginous hawk
  canvasback
* black tern
  Wilson's phalarope
  gadwall
* Nelson's sharp-tailed sparrow
  northern shoveler
  willet
* marsh wren
* long-billed curlew

Mountain plover, at risk

Sprague's pipit, at risk

Cockburn's lyrics remind us, "nothing worth having comes without some kind of fight." Yes, the trends are disheartening. Yes, the risks are real; but so is the potential for renewal. In spite of everything, native grassland still persists across the Great Plains, whether as isolated remnants in the Corn Belt or as vast sweeps of rangeland in the western provinces and states. And even though many of the prairie's special creatures are under severe stress, it is encouraging to remember that every single one of them is still with us. The only known exceptions are the passenger pigeon, a bird that was driven from superabundance into oblivion by market hunting in the nineteenth century, and the Rocky Mountain locust—the scourge of pioneer agriculture—which is thought to have vanished around 1900. A number of important subspecies have also been lost to the past, including the prairie-adapted races of wolves, grizzly bears, and bighorn sheep, each with its own genetic innovations.

Any human-caused extinction is a cause for regret. Yet given the intensity of development on the prairies in the last hundred-plus years, this record is better than anyone could have dared to hope for. It is a reminder that the grasslands and their native species are adaptable and tough, capable of coping with blizzards, fires, and hundred-year drought. Within limits, they can even cope with us. Yet natural adaptability on its own would not have been enough to bring the full complement of species through a century of crisis. Faced with rapid and violent change, several of the region's unique life-forms would almost certainly have disappeared without the intervention of conservation-minded people. The survival of species like the bison stands as proof that people who love the prairie and take a stand on its behalf are the last, best hope of the wild West.

Gone Today, Here Tomorrow

The story of environmental resistance on the Great Plains goes at least back to 1872 and the bloody era of the bison slaughter. In a year when 2 million bison were killed for their hides—their carcasses left to rot—a man named Samuel Walking Coyote took the exceptional step of rounding up seven orphaned calves and driving them from the Milk River country to his home on the Flathead Reservation in western Montana. There they and their burly descendants soon found their way into the hands of two local ranchers, Charles Allard and Michel Pablo, the latter a reformed buffalo hunter who wanted to make amends for his involvement in the bloodbath. Over the next twenty years, as the last of the wild herds were exterminated, Pablo and Allard devoted themselves to breeding and rearing their captive stock. By 1891, the two men possessed around thirty-five head, making theirs one of the largest herds of bison in existence. (The other survivors included a handful of stragglers in Yellowstone National Park, plus a couple of dozen private herds in Canada and the United States, with a combined population of two hundred to three hundred bison.)

Proving that conservation can be good business, Pablo and Allard managed their herd successfully for the next thirty years, selling hides and mounted heads and supplying live animals to zoos, parks, and other ranchers. Then, in the early 1900s, a change in government regulations opened the Flathead lands to homesteading—an intrusion that Pablo and Allard were helpless to resist—and there was no longer any place for them or their bison. Was the species finally being pushed into oblivion? Faced with this bleak prospect, a group of prominent eastern conservationists formed the American Bison Society, one of the first national environmental groups in North America, in 1905. Its objective was to establish wildlife refuges and, while there was still a chance, stock them with some of the captive bison. In short order, two small blocks of land were acquired, the Wichita Mountains National Wildlife Refuge in Oklahoma and the National Bison Range in Montana, the latter situated on what until then had been Flathead land, a maneuver that continues to be a bone of contention.

There are now about 500,000 bison on the Great Plains, up from a few hundred at the turn of the last century. Only a small fraction (about 4 percent) are managed for conservation. Small herd size and an inadequate land base frustrate efforts to maintain the species.

American burying
beetle, at risk

With the refuges in place, whether by fair means or foul, it was time to bring on the bison. A deal was struck with Pablo and Allard for the purchase of their herd, but Congress vetoed it as a waste of taxpayers' dollars. The Bison Society then set to work collecting private donations, in nickels and dimes, and ultimately raised enough funds to purchase a few head of breeding stock. Meanwhile, news that the Pablo-Allard herd was for sale sped north to Banff, Alberta, where it reached the ear of park superintendent Howard Douglas. A conservationist who already oversaw a small herd of bison in Rocky Mountain (now Banff) National Park, Douglas persuaded his superiors in the parks service to buy the Pablo-Allard outfit and transport the animals to Elk Island National Park, near Edmonton. Although it took six successive spring roundups to catch the half-wild beasts, just over seven hundred animals were eventually transferred across the border to relative safety.

Elk Island was supposed to be a stopover on the way to the animals' real home, the brand new Buffalo National Park near Wainwright, Alberta. About fifty of the animals managed to evade capture, however, and had to be left behind at Elk Island. The rest of the animals were duly transported to southern Alberta, where some of them were eventually crossbred with cows in an attempt to make them more amenable to domestication. Most of the bison in commercial production today carry cattle genes from this kind of misguided experiment. So, too, do many of the bison in Wood Buffalo National Park in northern Alberta, where the Wainwright herd—by then 6,700 strong—was transferred during World War II. The escapees at Elk Island, by contrast, are thought to breed true to their ancestral stock, as 100-percent-pure-and-unadulterated plains bison. (Other sources of pure plains-bison genetics include the herds at Wind Cave and Yellowstone national parks, in South Dakota and Wyoming, respectively, and likely those at Henry Mountains State Park in Utah, Sullys Hill National Game Preserve in North Dakota and Grand Teton National Park in Wyoming.) Of all the bison alive today, less than 1.5 percent are the Real Deal from a genetic point of view, and it is more by good luck than good management that we have any at all.

If there is a moral to this story, it is that prairie conservation is a high-stakes, high-risk undertaking, with a surprise at every bend and a constant call for new strategies and alliances. Plains bison would not have survived the last century without human help, however bumbling and misguided that "help" has sometimes been. The fact that the species exists today is a tribute to a fractious and imperfect collaboration between ranchers and city slickers, individuals and groups, Natives and non-Natives, citizens and civil servants, each group following its own bent—and with a little luck thrown in at critical moments. By engaging the problem from different angles, a diverse coalition of people came up with diverse approaches that, over time, have permitted them to confront a series of unexpected challenges. Just as an ecosystem relies on the interplay among species, so conservation depends on the interplay among people, with all their differences.

These days, the bigwigs in bison conservation are beginning to dream about boosting the recovery effort into a whole new dimension. In their minds' eyes, they can already envisage several herds of pure plains bison, each at least a thousand animals strong, roaming over wide vistas of windswept grassland. (Canadian Forces Base Suffield in Alberta, Charles M. Russell National Wildlife Refuge in Montana, and the Pine Ridge region of South Dakota and Nebraska are among the locations that have featured in these speculations.) Ideally, these populations would be free not only of cattle genes but also of cattle diseases, like anthrax and brucellosis, yet would be subject to a full suite of predators— perhaps grizzlies and certainly wolves—and other natural, evolutionary pressures. That's what it will take, the experts tell us, for plains bison to advance from mere survival toward restoration as the living, breathing embodiment of the prairie ecosystem, capable not only of restoring ecological function but also of "inspiring, sustaining and connecting human cultures." Is this just a pipe-dream? Or might this hopeful vision be just the spark we need to bring diverse interests together over a fresh pot of joe and get us talking to one another about the future?

Meanwhile, Back at the Ranch . . .

Bison carry the romance of the grasslands on their broad shoulders, so it is perhaps not surprising that they continue to inspire people to take action on their behalf. But many other grassland-adapted species have also benefited from human intervention. One thinks, for example, of the swift fox, a small, bat-eared, rodent-hunting canid that was once widely distributed across the Great Plains but that, after settlement, became severely diminished in both population and range. Thanks in large part to a captive-breeding program initiated by Beryl and Miles Smeeton of the Cochrane Ecological Institute, the species has been reintroduced to an area in southeastern Alberta and southwestern Saskatchewan from which it had disappeared, and a similar recovery is also under way on the Blackfeet reservation in Montana. Meanwhile, reintroductions are in progress on Kainai (Blood tribal) lands in Alberta and at Badlands National Park and the Bad River Ranches in South Dakota. At the same time, a host of agencies and individuals are engaged in the conservation of other organisms, from endangered prairie orchids to native butterflies and bees to rattlesnakes and other reptiles that rely on native prairie.

"To keep every cog and wheel," the American environmental philosopher Aldo Leopold once wrote, "is the first precaution of intelligent tinkering." But simply retaining all the pieces of the ecosystem will not be enough in itself; we also need places where the cogs and wheels can be put back together. To ensure their long-term survival, grassland species need wild grasslands—broad expanses of native prairie that, through a natural process of disturbance and renewal, are able to maintain a living mosaic of habitats for a full complement of birds and animals. Despite all the historic losses, large, areas of more-or-less natural prairie still exist, especially on the rangelands of the northwestern short- and mixed-grass ecoregions in Alberta, Saskatchewan, Montana, Wyoming, the Dakotas, and Nebraska. To this day, there are at least ten regional landscapes on the northern plains that are dominated, as far as the eye can see, by expanses of native prairie. Several of these areas encompass as much as 4,600 square miles (12,000 square kilometers), larger than either Jasper or Yellowstone national parks, making them of real significance for conservation. With this in mind, the World Wildlife Fund recently identified the northern plains as a "biologically outstanding" habitat and one of its Global 200 targets for biodiversity conservation. Opportunities for large-scale conservation on the southern plains are also coming into focus.

Very little of this land has been formally protected by law. In fact, across the Great Plains as a whole only about 1 percent of the countryside has been set aside in parks or conservation reserves, less than in any other biome in North America. The surviving expanses of native prairie are hard-working landscapes that today, as in the past, provide the basis for western cattle production. This is cowboy country. Ranches are not wildlife refuges, and over the years, ranchers have made it clear that varmints like prairie dogs, wolves, and wild bison won't find a warm welcome here. But at the same time, ranching has placed a value on both wild prairie as grazing land and on the esthetics of broad horizons. For many ranchers, maintaining large expanses of native pasture in productive condition has been a labor of love, as well as an act of economic self-interest. It is a tribute to their efforts that several recent conservation projects, including

Blowout penstemon, at risk

Western prairie fringed orchid, at risk

> **WIDE OPEN FOR CONSERVATION**

In 2004, the Commission for Environmental Cooperation and the Nature Conservancy brought together a panel of experts to identify the most important and promising sites for grassland conservation in North America.

The regions they identified, as shown on this map, represent the best of what's left, combining high biodiversity value with expanses of intact grassland. Similar opportunities also exist in three areas of northern Mexico: Tokio-Mapimi, Marfa-Big Bend-Maderas del Carmen, and the Sierra Madre Occidental Foothills.

 Central Plains

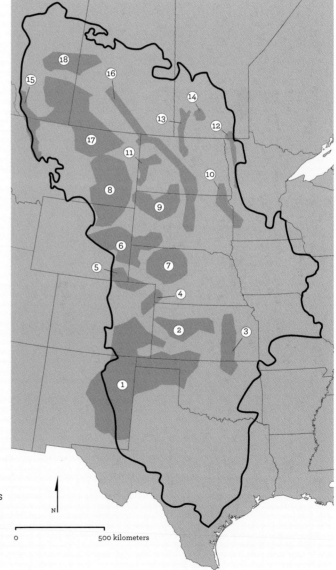 Grassland Ecologically Significant Regions

Grassland Ecologically Significant Regions

1. Southern Prairie
2. Smoky Hills
3. Flint Hills
4. Arickaree
5. Pawnee Grasslands
6. Thunder Basin—Conata
7. Sandhills
8. Powder River—Big Open
9. Cheyenne Prairie
10. Prairie Coteau—Sheyenne Delta
11. Little Missouri
12. Agassiz Tallgrass
13. Manitoba Mixed-grass—Towner Sandhil
14. Carberry Sandhills
15. Rocky Mountain Front
16. Missouri Coteau
17. Medicine Line Mixed-grass
18. Northern Fescue

POTENTIAL AREAS FOR LARGE-LANDSCAPE CONSERVATION ON THE NORTHERN GREAT PLAINS

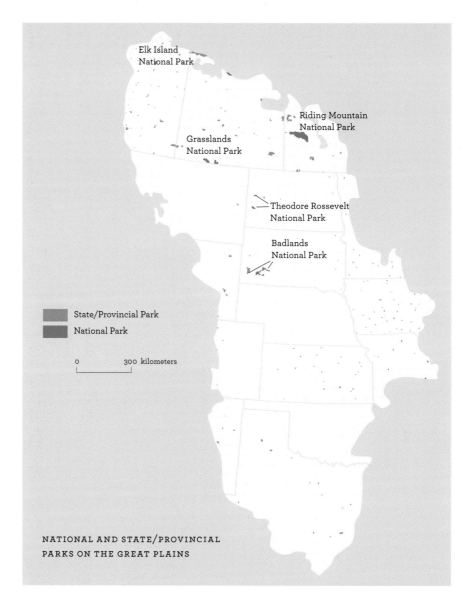

Elk Island
National Park

Riding Mountain
National Park

Grasslands
National Park

Theodore Rossevelt
National Park

Badlands
National Park

State/Provincial Park

National Park

0   300 kilometers

NATIONAL AND STATE/PROVINCIAL
PARKS ON THE GREAT PLAINS

Grasslands National Park and Old Man on His Back Conservation Area in Saskatchewan and the Tall Grass Prairie National Preserves in Oklahoma and Kansas, have been established on lands that were previously managed as commercial ranches.

Ranching not only has the advantage of preventing the prairie from being plowed up. It also helps to keep it from becoming fragmented. Anything that takes a bite out of the natural grassland, whether it be a tame pasture, a pricey ranchette, or an oil field with its network of service roads, breaks up the landscape and subtly alters its ecological function. A study conducted in Colorado

COWBOY CONSERVATION

If there were legal protection for endangered societies as well as species, the ranching communities of the western plains would surely top the list. In one struggling town after another, school enrollments are declining, storefronts are boarded up, and the ranchers who meet on coffee row are no longer young. A uniquely western way of life is at risk of blinking out.

In the past, conservationists and ranchers have often had a testy relationship, as insensitivity to local values met head on with local suspicion of outsiders. But tempers have mellowed with time, and the old adversaries are increasingly able to find common ground in their concern for the future. Ranchers want to earn a living from ranching, and conservationists want them to succeed, as part of a strengthened and diversified nature-based economy.

In a report entitled *New Directions for the Prairie Economy: Connecting Conservation and Rural Development in the Northern Great Plains,* published in 2009, the World Wildlife Fund presents a twelve-point plan for "fostering nature-based economic development in ways that also support the goals of biodiversity conservation." Key recommendations include: expanding ranch-based ecotourism (including hunting) to "reward landowners who conserve biodiversity"; developing markets for biodiversity-friendly beef and other grassland products, possibly through certification and labeling; enlarging ecological reserves to encompass at least 10 percent of the northern plains; expanding support for Indigenous people's ecological and cultural aspirations; securing adequate funding for existing conservation programs, which are chronically underresourced; and providing compensation to ranchers for the billions of dollars worth of ecological services they provide, including protection of watersheds, prevention of soil erosion, and the sequestration of carbon.

"We have no illusions that a nature-based economy is the main path to meeting the region's rural development needs," the report's authors admit, "or that rural development is the best path to conserving the region's biodiversity. But we do believe there is a place where these two paths intersect, where there is common ground for merging the interests and needs of both for mutual benefit."

showed that the subdivision of ranch land into acreages led to the displacement of grassland birds, such as lark buntings and meadowlarks, and their replacement with robins, magpies, and other common-and-garden species. The pattern was the same for carnivores, with coyotes and bobcats ceding their role as predators to domestic pets. Most worrying of all was the discovery that these cut-up tracts of prairie were much more susceptible than working ranch lands to intrusions by a long list of invasive plants, including introduced species such as smooth brome, crested wheatgrass, leafy spurge, and dozens of others. Given an inch of bare ground, these aggressive weeds will literally take a mile, eventually overwhelming and choking out the natural vegetation. The more lines of disturbance are scratched across the country, the more access these invaders gain and the more quickly the native prairie is forced to give way.

There are more than 460 "species of concern" on the Great Plains. These are organisms faring so badly in population and reproductive success that they are feared to be heading toward extinction. Although compilations vary, the list includes something like 18 species of crayfish and other crustaceans, 19 species of reptiles (among them the Black Hills redbelly snake), 19 amphibians (including a kind of blind salamander found only in Texas caves), 21 birds (the whooping crane, for example, and the lesser prairie chicken), 26 mammals (including the black-footed ferret and the swift fox), 33 fishes (sturgeon, minnows, darters, shiners, and chubs), 41 snails and mussels, 57 spiders and insects (including the so-called superb grasshopper, also from Texas), and well over 200 kinds of plants (the blowout penstemon, for example, found only on active sand dunes, and the delicate prairie fringed orchids, among many others).

One species that is conspicuously absent from this disheartening roster is the black-tailed prairie dog, though it also deserves mention. After initially declaring the species "endangered but precluded" from listing because of other priorities, the U.S. Fish and Wildlife Service changed its mind in 2007. The prairie dog is no longer a candidate for listing. Meanwhile, their plight continues to affect dozens of other species that are specialized to live in and around prairie-dog towns, including such vulnerable birds as the mountain plover, ferruginous hawk, and burrowing owl.

Species that are habitat specialists, rather than generalists, are at elevated risk of extinction because they are not well equipped for change. If their habitat vanishes, they are finished. It is therefore not surprising that most of the species at risk, about 70 percent, are endemics. These are organisms with unique adaptations to unique habitats. The narrower the specialization, the more vulnerable a species is to becoming rare and ultimately going extinct. For this reason, the Edwards Plateau of Texas, which has more than its share of unusual organisms, is also a zone of endangerment.

Monarch butterfly, at risk

One way to prevent these losses is to keep livestock on the range and to maintain the historical use of the land for grazing. Perhaps the best-known proponent of this approach to conservation is billionaire-businessman-turned-cowpoke Ted Turner. In an attempt to disprove the adage that the best way to make a small fortune in the livestock industry is to start with a large one, Turner has invested a portion of his wealth in a string of ranches across the Great Plains, with the intention of showing the world that conservation can pay. Now the largest bison producer in the country, with a combined herd on his various holdings of some forty thousand to fifty thousand head, he makes it a policy to stock his ranches below industry standards, both as a means of saving money (no need to provide extra feed during droughts) and as a safeguard against overgrazing. What's more, through the Turner Endangered Species Fund, he and his family are attempting to improve the prospects of prairie dogs, black-footed ferrets, and other species at risk both on the Turner properties and elsewhere. And finally, he has launched a chain of restaurants under the banner of Ted's Montana Grill, where diners can share his taste for prairie conservation.

The livestock industry has been the subject of intense criticism from environmentalists, often for valid reasons. But not all ranches are created equal. Under thoughtful management, working rangelands can be highly productive for wildlife and, at their best, have been known to support more native species, at higher densities, than are found on nearby wildlife refuges. Buying range-fed beef or bison from a conservation-minded producer has a significance that extends far beyond the dinner table. Happily, these prairie-friendly products are readily available in many parts of the Great Plains, whether purchased at the farm gate or through farmers' markets, health food stores, and other alternative outlets. A number of producers are also offering their range-fed meats to a broad public through the Internet, under trademarks such as Conservation Beef and Wild Idea bison.

Two other strategies for conserving native grassland deserve mention. The first is a type of legal contract known as a conservation easement. Simply put, an easement is a commitment made by a landowner to exempt a specified parcel of land from future development. In return, he or she qualifies to receive compensation from the sponsoring organization, usually an environmental group (like the Nature Conservancy or Ducks Unlimited) or a government agency (like the United States Fish and Wildlife Service). Although the terms of the contracts vary, the restrictions typically include a ban on cultivation, subdivision, and the destruction or degradation of wetlands; sometimes energy

Conserving rangelands ensures a future for species like the prairie rattlesnake. To survive, rattlers need grasslands for summer hunting and bankside crevices as hibernacula, or communal winter dens. Though venomous, prairie rattlesnakes are not aggressive and will typically slither away if given half a chance.

development and road building are also prohibited. These limitations apply not only to the current owner but also to anyone who subsequently acquires the land, thereby achieving permanent protection at the cost of a onetime payment. And even though the payouts are modest, they have been welcomed by ranchers, who find themselves chronically caught between high costs of production and low prices at the livestock market.

A second strategy—setting aside sweeping landscapes for wildlife, in parks and ecological reserves—understandably raises the hackles of private landowners. But with fair compensation for those who choose to sell and the promise of a diversified economy, built around ecotourism, to sustain those who remain on the land, there can sometimes be a meeting of minds, a softening of differences, and a merging of agendas.

Growing Conservation

The importance of protecting the surviving horizons of native grassland can scarcely be overstated. But it is not the only urgent priority for prairie conservation. To borrow again from Aldo Leopold, relegating wild prairie to the western rangelands is like relegating happiness to heaven; one may never get there to enjoy it. Meanwhile, there are other pleasures. The parts of the country that have been altered by farming still shimmer with life. Roadsides are heady with wild roses; geese explode out of farmers' fields; pretty little deer mice patter around barns and granaries. Although farmland is not prairie, it nonetheless provides habitat for many species of native animals and plants. Sometimes these farmyard populations are all that is left, remnants of a world that has otherwise vanished.

Take the case of the tallgrass prairie. Apart from a relatively small enclave in the Flint Hills and Osage Plains of Kansas and Oklahoma (where the soil is too rocky to till), this ecoregion has been almost 100 percent converted to crop production. We're talking about the row-after-row-after-row-ness of the Corn Belt. All that remains of the native vegetation are small, isolated patches that somehow escaped the plow, leaving a tantalizing glimpse of a country alight with butterflies and bright with flowers. Although some of these fragments are smaller than backyard gardens, each one demands attention and care as a unique example of a critically imperiled ecosystem. (Because of local variations in growing conditions and through pure chance, no two remnants have exactly the same species in the same proportions.) Through the combined initiatives of national and local organizations, many of these prairie remnants are now protected by law and are intensively managed to prevent the incursion of woody invasives and other takeover artists. Similar rescue efforts have also begun in the mixed-grass ecoregions.

These islands of survival provide critical habitat for many species at risk: the dickcissel, the regal fritillary butterfly, the ornate box turtle, the prairie rattlesnake, and on and on. But, alas, these small, isolated populations remain under constant threat. If some misfortune befalls them—whether through disease or predation or drought—there are no neighboring populations to move in and replace them, with the result that a single disastrous season could wipe them out. The only possible solution is, wherever possible, to create blocks of habitat large enough to support viable populations or to provide corridors between the existing fragments. Hence, the current lively interest in prairie restoration.

Regal fritillary
butterfly, at risk

Plains wolf, extirpated

Going against the historical trend of plowing prairie up, conservation-minded people have begun to replant it. This has called for innovation in both equipment and techniques and has sparked the development of a native-seed industry. Creating new prairie is tricky, expensive and, in terms of species richness, never a complete success, but it is an inspiring step in the right direction.

The cause of prairie restoration has found some unexpected advocates, among them the Iowa Department of Transportation. Iowa is farming country taken to the extreme, with only scant vestiges of native prairie. What the state does have, however, is a go-anywhere grid of roads, all of which have vegetated verges. Taken together, these strips add up to about half a million acres (roughly 2,000 square kilometers) of unproductive land that requires mowing, spraying, and other regular maintenance. In an attempt to reduce costs in the late 1980s, the transportation authorities began to experiment with the use of native plants, on the assumption that they were adapted to local conditions and could look after themselves. Since then, more than 50,000 acres (20,000 hectares) of roadside have been seeded, a little more every year, to either a four-grass mixture—typically big and little bluestem, sideoats grama, and Indian grass—or to a colorful assortment of native grasses and wildflowers.

The results have exceeded all expectations. In addition to controlling expenses, the flower-rich plantings in particular have become slender oases of life, blooming not only with flowers but also with butterflies. In one case, for example, researchers found five times as many butterflies and twice as many species in the high-quality restorations as in comparable grassy or weedy ditches. This success has inspired the Iowa Transportation Commission to pump millions of dollars into the Living Roadways Program, making the state a leader in what it calls "Eco-Logical transportation."

Big bluestem pushes up through a profusion of goldenrod and other wild flowers in a prairie-restoration project in western Illinois.

The New Green Revolution

The thought that corridors of prairie might one day run along roads and highways across the Great Plains is cheerful to contemplate. But even if it eventually happens, it will not be enough: life cannot be relegated to the margins. In areas devoted to farming, the working landscape is an ecoregion all its own, where there is a great deal more at stake than crop production. Admittedly, cultivated land provides less-than-ideal habitat for many organisms, notably for those that cannot cope with chronic disruption. Yet for species that have been able to make the transition to the agroecosystem, farmland is home, and what happens there is critical to their survival.

The acceptance of conservation as an everyday aspect of farming practice has been halting. On the plus side, thanks to the object lessons of the Dirty Thirties and the 1980s drought, the importance of soil conservation is now well established and has been widely translated into both policy and action. On the debit side, however, conservation of wildlife habitat is often a hard sell. Sometimes producers resist as a matter of principle. Having devoted their efforts to producing food to feed a hungry world, they argue that leaving space "idle" for wildlife is a misuse of productive potential. But while this position would have been irresistible in the 1960s, when an exploding human population faced a net shortage of food, it is no longer convincing. In the intervening decades, a complex suite of developments, some of them halfway around the world, have opened up new and more hopeful options for prairie agriculture. The achievements of the world's farmers since World War II have been stunning. Between 1950 and 1992, for example, world grain production increased by 170 percent, with only a 1 percent expansion of the area under cultivation. Although many people still go hungry, there is more than enough food available to feed everyone on Earth, a situation that is expected to persist well into the future. This triumph has come about as the result of a no-holds-barred commitment to maximizing production, known colloquially as the Green Revolution. Sparked by an Iowa-born scientist named Norman Borlaug, who won a Nobel Prize in 1970 for his work, the revolution was based on the development of high-yield varieties of wheat and corn. And yield they did, provided that they were supplied with ample stores of nitrogen, from artificial fertilizer, and abundant water, typically from irrigation.

Unfortunately, the environmental costs of this high-tech fix are only now being seen. On the Great Plains, in particular, the damage has included the depletion of aquifers through irrigation, the poisoning of groundwater with agricultural chemicals, and the overfertilization of entire river systems. There is now an enormous dead zone in the Gulf of Mexico, caused by an influx of nutrients, mostly nitrogen and phosphorus, from artificial fertilizers applied to farmlands in the Missouri/Mississippi drainage.

But perhaps the most serious and unexpected downside of the Green Revolution was the collapse of world markets. As the supply of farm commodities

Ferruginous hawk, at risk

outstripped demand, prices dropped to levels that hadn't been seen since the Great Depression. In response to this economic disaster, many producers rooted out fencerows, shelterbelts, wetlands, and other uncropped lands in an attempt to increase their salable harvest. And as if pressure from the marketplace were not enough, farmers often received extra inducement from government income-support programs, in which entitlements were based on the area under cultivation. The more you plowed up, the more you stood to receive. The result of these combined forces has been an incremental loss of habitat from the farm landscape and a corresponding decline in the abundance and diversity of wildlife.

Plus ça change, plus c'est la même chose. Since about 2006, world grain prices have come out of their decades-long doldrums and begun spiking up, doubling, tripling, and quadrupling to record-high levels. Although partly driven by global consumer demand, the price rise has really taken off because of the rocketing market for biofuels. Crops like wheat and corn, once valued as foodstuffs, have now become feedstock for ethanol production. While the biofuel boom has provided welcome relief for farmers (especially the big players in the game), it has done nothing to ease the pressure on prairie landscapes. In fact, the drive toward industrial intensification has itself intensified. In the U.S., for example, thousands—perhaps even millions—of acres of marginal farmland that were seeded to grass under the Conservation Reserve Program in the 1980s and 1990s are being brought back under the plow, at an untold cost to grassland birds and waterfowl. Meanwhile, carbon that had been sequestered in the soil is being released when the land is worked, putting the lie to biofuel's promise of a low-carbon future.

The grain-based ethanol industry has developed as an instrument of public policy, backed by millions in public funds. Fortunately, public influence is also being exerted in less ambiguous directions. The last decade has seen the creation of innovative programs (typically underfunded but important nonetheless) that pay farmers for managing their private land for the public good by providing services to the environment. Both the 2008 Agricultural Policy Framework in Canada and the 2009 Farm Bill in the U.S. provide limited compensation to farmers who fulfill specific agreements to reduce effluents and emissions, conserve soil fertility, provide habitat for wildlife, or provide other benefits to society.

Of course, long before these programs existed, there were already farmers who set high standards for themselves and managed their lands for

Concho water snake,
at risk

Texas blind salamander,
at risk

Neosho madtom,
at risk

conservation. For some, it was a simple matter of taking a little extra time to maneuver around marshes and ponds or delaying their haying operations until late in the season when the ground-nesting birds are gone. But others have made a day-in, day-out commitment to farming *with*, rather than on, the land, using methods that attempt to mimic the natural ecosystem. Sometimes disparaged as hopeless romantics with a nostalgic attachment to the past, these organic, or holistic, producers are practitioners of a high-yield, knowledge-based, thoroughly modern system that continues to evolve as agroecological science advances. Drawing on research and personal experience, organic farmers use a variety of techniques, such as green manures (crops that are grown and then plowed under to build the soil), intercropping (green manures and crops sown together), and crop rotations (a sequence of crops planted in a field in succeeding years), to add diversity and complexity to their fields. New methods for controlling weeds without tillage, by using cover crops as mulch, are the focus of intensive investigation. Although not a panacea, organic techniques achieve many well-documented benefits, including improved retention of organic matter in the soil, increased diversity of soil organisms, fewer plant diseases and pest infestations, more variety in cropping systems, reduced run-off of inorganic nutrients, and a lower burden of environmental toxins. Despite the inevitable disturbance to wildlife caused by field work, organic farms also generally support more species in greater abundance—everything from spiders to bees to birds—than are found on their conventional counterparts.

Yields from organic fields are often lower than those achieved on other farms, and organics are sometimes derided as boutique agriculture. But researchers in South Dakota have found that organic methods can be highly productive, especially during droughts, with yields that equal or even outstrip those of high-input farms. And a recent analysis by scientists from the University of Michigan suggests that organic farming has the potential to meet the caloric requirements of a hungry world, without increasing in the area under cultivation and at a reduced cost to soil, water, air quality, and life in general.

Managing for Wildness

All of these complexities—of rangeland and farmland, easements and frameworks, opportunities and options—can be summed up in two basic concepts. They are the mantras of prairie conservation. The first is *to protect and enhance wild prairie* wherever it still exists, whether as large, connected landscapes or, where no alternative is left, as one-of-a-kind fragments. The second, often overlapping priority is *to manage the working landscape for wildness* so that it not only serves the interests of people but also supports a diversity of swimming, flying, walking, and crawling forms of life. Achieving these goals will not be easy. Failing to achieve them will mean a continuing downward trend for many of the prairie region's unique ecoregions and species.

These priorities take on even greater urgency in the context of climate change. The grasslands as we know them emerged thousands of years ago, at the end of a three-million-year-long ordeal of glaciation. The difference between the desolation of the Ice Age and the birth of the prairies was a natural warming trend that caused the average global temperature to rise by about 9°F (5°C). Now, the experts tell us, we are about to experience a perturbation of similar magnitude but one that we ourselves have triggered. The problem, of course, is the thick, insulating blanket of carbon dioxide, nitrous oxide, and other industrial emissions that envelops the Earth and prevents heat from radiating away from the surface. Depending in part on how successful we are in curbing this process, the average world temperature is expected to increase by

Protection and enhancement of the prairie's uniquely productive wetlands is a top priority everywhere on the Great Plains.

A female mule deer keeps watch over her investment in the future.

somewhere between 2.0°F and 11.54°F (1.1°C and 6.4°C) before 2100. If these predictions are even close to being accurate, the Earth may soon be hotter than at any time in the past million years, and the change will have occurred more rapidly than any on record.

Despite the clamor of dissent in the media and the blogosphere, climatologists are increasingly united about the probable, large-scale consequences of the greenhouse effect. In addition to atmospheric warming, the outlook includes the melting of polar ice packs, a rise in sea levels (with consequent flooding), and an increase in natural disasters such as droughts, fires, hurricanes, and tornadoes. But although the big picture is clear—even distressingly so— attempts to produce localized forecasts have so far been far less clear-cut. In the case of the Great Plains, for example, the predictions call for the climate to be hotter and more extreme than it has been in the past, with lower water levels in rivers and lakes and a reduced area of wetlands. But nobody knows what will happen.

The wild prairies are, in the deepest sense, a manifestation of the climate. From the ground up, the living world is attuned to wind and rain, sun and snow, seasons of death and seasons of growth. As these basic realities are altered, everything will be touched, and change will ripple and ricochet through the ecosystem. On the one hand, human land-use patterns are certain to be transformed, as people attempt to adapt to a rapid-fire succession of opportunities and challenges. Whether these shifts in human activity will be good or bad for wildlife is anyone's guess. At the same time, the changed climatic regime will also affect wildlife directly, by opening up new prospects for some species (especially generalists) and closing in on others (particularly those that are isolated or have specialized requirements). Are we heading toward a nightmarish future dominated by weeds and pests, in which the prairies are stripped of their special beauty and begin to look like everywhere else?

There is no way to hold back the future. But we can shape the course of events by engaging—fully, deeply, and passionately—with the present. The survival of the wild prairie and its creatures will depend, in no small part, on our ability to ensure their well-being right now. By protecting and enhancing wild prairie and managing the working landscape for wildness, we can strengthen and enhance the ecosystem, in all its diversity and abundance, both for our own sake and for those who come after us. This approach is sometimes referred to as a strategy of "no regrets," because the work is worth doing now, no matter what happens next.

THE GREAT PLAINS

{ FOR MORE INFORMATION }

A complete catalog of the written sources that were consulted in the preparation of this book would fill a volume of its own. This list is therefore selective and is primarily intended to provide a starting point for further study and engagement. Your bookseller will be able to direct you to references that are appropriate for your area, and local conservation groups can guide you to opportunities to learn and participate.

CONSERVATION CONNECTIONS: PLACES TO BEGIN

American Prairie Foundation, www.americanprairie.org/
Biodiversity Conservation Alliance, www.voiceforthewild.org/
Great Plains Restoration Council, www.gprc.org/
Intertribal Bison Cooperative, www.itbcbison.com/index.php
Nature Conservancy of Canada, www.natureconservancy.ca/
The Nature Conservancy, www.nature.org/
World Wildlife Fund, www.worldwildlife.org/what/wherewework/ngp/

LOCAL GUIDES/GENERAL REFERENCES: A SAMPLER

Acorn, John. *Bugs of Alberta*. Edmonton: Lone Pine, 2000.
———. *Butterflies of Alberta*. Edmonton: Lone Pine, 1993.
———. *Tiger Beetles of Alberta*. Edmonton: University of Alberta Press, 2001.
American Ornithologists' Union. *The Birds of North America: Life Histories for the 21st Century*. Washington, D.C., 1992 onwards. A series of authoritative monographs.
Brown, Annora. *Old Man's Garden*. Sidney: Gray's, 1970.
Brown, Lauren. *Grasslands: National Audubon Society Nature Guide*. New York: Knopf, 1998.
Costello, David. *The Prairie World*. New York: Thomas Crowell, 1969.

Davis, William B., and David J. Schmidly. *The Mammals of Texas*. Texas Parks and Wildlife Department, 1994. www.nsrl.ttu.edu/tmot1/.
Foresman, Kerry R. *The Wild Mammals of Montana*. Special Publication No. 12. American Society of Mammalogists, 2001.
Grasshoppers of Wyoming and the West. www.uwyo.edu/grasshoppersupport/Html_pages/ghwywfrm.htm.
Handbook of Texas Online. Information on a wide variety of subjects. www.tshaonline.org/handbook/online.
Hoberg, Ted, and Cully Gause. "Reptiles and Amphibians of North Dakota." *North Dakota Outdoors* 55 (1992): 7–19. www.npwrc.usgs.gov/resource/herps/amrepnd/index.htm.
Johnsgard, Paul A. *Grassland Grouse and Their Conservation*. Washington, D.C.: Smithsonian, 2002.
———. *The Nature of Nebraska*. Lincoln: University of Nebraska Press, 2001.
———. *Prairie Birds: Fragile Splendor in the Great Plains*. Lawrence: University Press of Kansas, 2001.
———. *This Fragile Land: A Natural History of the Nebraska Sandhills*. Lincoln: University of Nebraska Press, 1995.
Jones, J. Knox. *Mammals of the Northern Great Plains*. Lincoln: University of Nebraska Press, 1983.
Kindscher, Kelly. *Edible Wild Plants of the Prairie*. Lawrence: University Press of Kansas, 1987.
Krausman, Paul R., ed. *Rangeland Wildlife*. Denver: Society of Range Management, 1996.
Ladd, Doug. *Tallgrass Prairie Wildflowers: A Falcon Field Guide*. Helena: Falcon, 1995.
Laurenroth, William K., and Ingrid C. Burke. *Ecology of the Shortgrass Steppe: A Long-Term Perspective*. Oxford: University of Oxford Press, 2008.
Madson, John. *Where the Sky Began: Land of the Tallgrass Prairie*. Boston: Houghton Mifflin, 1982.

Northern Prairie Wildlife Research Center. Information on a wide range of prairie organisms and issues. www.npwrc.usgs.gov/.

Preston, William B. *The Amphibians and Reptiles of Manitoba.* Winnipeg: Manitoba Museum of Man and Nature, 1982.

Raventon, E. *Island in the Plains: A Black Hills Natural History.* Boulder: Johnson, 1994.

Reichman, O.J. *Konza Prairie: A Tallgrass Natural History.* Lawrence: University Press of Kansas, 1987.

Russell, Anthony P., and Aaron M. Bauer. *Amphibians and Reptiles of Alberta: A Field Guide and Primer of Boreal Herpetology.* Calgary: University of Calgary Press, 2000.

Saskatchewan Environment and Resource Management. *Natural Neighbours: Selected Mammals of Saskatchewan.* Regina: Canadian Plains Research Center, 2001.

Vance, F.R., et al. *Wildflowers Across the Prairies.* Vancouver: Greystone, 1999.

CHAPTER 1: WHERE IS HERE?

Alt, David, and Donald W. Hyndman. *Roadside Geology of Montana.* Missoula: Mountain Press, 1986.

Boal, Frederick W., and Stephen A. Royle, eds. *North America: A Geographical Mosaic.* London: Arnold, 1999.

Bolen, Eric G. *Ecology of North America.* New York: John Wiley and Sons, 1998.

Bragg, Thomas B. "The Physical Environment of the Great Plains Grasslands." In *The Changing Prairie*, edited by Anthony Joern and Kathleen H. Keeler, 49–81. New York: Oxford University Press, 1995.

Brown, Robert H. *Wyoming: A Geography.* Boulder: Westview Press, 1980.

Fenneman, Nevin M. *Physiography of Eastern United States.* New York: McGraw-Hill, 1931.

———. *Physiography of Western United States.* New York: McGraw-Hill, 1931.

Fung, Kai-lu, ed. *Atlas of Saskatchewan.* Saskatoon: University of Saskatchewan, 1999.

Griffiths, Meland, and Lynnel Rubright. *Colorado: A Geography.* Boulder: Westview, 1983.

Hart, Richard H., and James A. Hart. "Rangelands of the Great Plains Before European Settlement." *Rangelands* 19 (1997): 11.

Holliday, Vance T., et al. "The Central Lowlands and Great Plains." In *The Physical Geography of North America*, edited by Antony R. Orme. New York: Oxford University Press, 2002.

Jordan, Terry G., et al. *Texas: A Geography.* Boulder: Westview, 1984.

McKnight, Tom L. *Regional Geography of the United States and Canada.* Upper Saddle River: Prentice Hall, 1997.

Nelson, H.L. *A Geography of Iowa.* Lincoln: University of Nebraska Press, 1968.

Ostlie, W.R., et al. *The Status of Biodiversity in the Great Plains.* The Nature Conservancy, Arlington: 1997.

Radenbaugh, Todd A., and Patrick Douaud, eds. *Changing Prairie Landscapes.* Regina: Canadian Plains Research Center, 2000.

Rafferty, Milton D. *Missouri: A Geography.* Boulder: Westview, 1983.

Ricketts, Taylor H., et al. *Terrestrial Ecoregions of North America: A Conservation Assessment.* Washington, D.C.: Island Press, 1999.

Sims, Phillip L., and Paul G. Risser. "Grasslands." In *North American Terrestrial Vegetation*, edited by Michael G. Barbour and William Dwight Billings, 324–56. Cambridge: Cambridge University Press, 2000.

Trenhaile, Alan S. *Geomorphology: A Canadian Perspective.* New York: Oxford University Press, 1998.

Weaver, J.E., and F.W. Albertson. *Grasslands of the Great Plains: Their Nature and Use.* Lincoln: Johnsen, 1956.

World Resources Institute—PAGE, 2000. "Global Extent of Grasslands." http://earthtrends.wri.org/text/forests-grasslands-drylands/map-229.html.

CHAPTER 2: DIGGING INTO THE PAST

Anderson, Wayne I. *Geology of Iowa: Over Two Billion Years of Change.* Ames: Iowa State University Press, 1983.

Beaty, Chester B. *The Landscapes of Southern Alberta: A Regional Geomorphology.* Lethbridge: University of Lethbridge, 1975.

Belcher, C.M. "Impacts and Wildfire: An Analysis of the K-T Event." In *Biological Processes Associated with Impact Events*, edited by Charles Cockell, et al., 221–43. Berlin: Springer, 2006.

Beveridge, Thomas R. *Geologic Wonders and Curiosities of Missouri*. Rolla: Missouri Division of Geology and Land Survey, 1980.

Bluemle, John P. *The Face of North Dakota: The Geologic Story*. North Dakota Geological Survey Education Series 11, 1991.

Buchanan, Rex, ed. *Kansas Geology: An Introduction to Landscapes, Rocks, Minerals, and Fossils*. Lawrence: University Press of Kansas, 1984.

Chin, Karen, et al. "A King-sized Theropod Coprolite." *Nature* 393 (1998): 680–82.

Chronic, Halka. *Roadside Geology of Colorado*. Missoula: Mountain Press, 1980.

Dort, Wakefield, Jr., and J. Knox Jones, Jr. *Pleistocene and Recent Environments of the Central Great Plains*. Lawrence: University Press of Kansas, 1972.

Gordon, Anthony. *Geology of Saskatchewan: A Historical Approach*. Regina: Western Extension College Educational Publishers, 1979.

Gould, Stephen Jay. *Wonderful Life: The Burgess Shale and the Nature of History*. New York: W.W. Norton, 1989.

Gries, John Paul. *Roadside Geology of South Dakota*. Missoula: Mountain Press, 1996.

Lageson, David R., and Darwin R. Spearing. *Roadside Geology of Wyoming*. Missoula: Mountain Press, 1988.

Maher, Harmon D. *Roadside Geology of Nebraska*. Missoula: Mountain Press, 2003.

Ojakangas, Richard W., and Charles L. Matsch. *Minnesota's Geology*. Minneapolis: University of Minnesota Press, 1982.

Pielou, E.C. *After the Ice Age: The Return of Life to Glaciated North America*. Chicago: University of Chicago Press, 1991.

Russell, Dale A. *An Odyssey in Time: The Dinosaurs of North America*. Toronto: University of Toronto Press, 1989.

———. *A Vanished World: The Dinosaurs of Western Canada*. Natural History Series No. 4. National Museums of Canada, 1977.

Skwara, T. *Old Bones and Serpent Stones: A Guide to Interpreted Fossil Localities in Canada and the United States*. Vol. 2, *Western Sites*. Blacksburg: McDonald and Woodward, 1992.

Spearing, Darwin. *Roadside Geology of Texas*. Missoula: Mountain Press, 1979.

Storer, John. *Geological History of Saskatchewan*. Regina: Saskatchewan Museum of Natural History, 1989.

Trimble, Donald E. *The Geologic Story of the Great Plains*. Geological Survey Bulletin 1493, 1980. www.lib.ndsu.nodak.edu/govdocs/text/greatplains/text.html.

Troeger, Jack Clayton. *From Rift to Drift: Iowa's Story in Stone*. Ames: Iowa State University Press, 1983.

CHAPTER 3: THE GEOGRAPHY OF GRASS

Axelrod, Daniel I. "Rise of the Grassland Biome, Central North America." *Botanical Review* 51 (1985): 164–201.

Borchert, John R. "The Climate of the Central North American Grassland." *Annals of the Association of American Geographers* 40 (1950): 1–39.

Bryson, Reid A., and F. Kenneth Hare, eds. *Climates of North America*. New York: Elsevier, 1974.

Collins, Scott L., and Linda L. Wallace, eds. *Fire in North American Tallgrass Prairies*. Norman: University of Oklahoma Press, 1990.

Dey, B. "Nature and Possible Causes of Droughts on the Canadian Prairies—Case Studies." *Journal of Climatology* 2 (1982): 233–49.

Harrington, John A., Jr., and Jay A. Harman. "Climate and Vegetation in Central North America: Natural Patterns and Human Alterations." *Great Plains Quarterly* 11 (Spring 1991): 103–12.

Herring, David. "Dry Times in North America." NASA Earth Observatory, 2000. http://earthobservatory.nasa.gov/Features/NAmerDrought/.

Joern, Anthony, and Kathleen H. Keeler, eds. *The Changing Prairie: North American Grasslands*. New York: Oxford University Press, 1995.

Johnson, S.R., and Aziz Bouzaher, eds. *Conservation of Great Plains Ecosystems: Currrent Science, Future Options*. Dordrecht: Kluwer Academic Publishers, 1995.

Laird, Kathleen R., et al. "Greater Drought Intensity and Frequency Before ad 1200 in the Northern Great Plains, USA." *Nature* 384 (1996): 552–54.

McNaughton, S.J., et al. "Interactive Processes in Grassland Ecosystems." In *Grasses and Grasslands*, edited by C. Barnard. New York: St. Martin's Press, 1964.

Nichols, Douglas J., and Kirk R. Johnson. *Plants and the K-T Boundary*. Cambridge: Cambridge University Press, 2008.

Rissler, Paul G. "Grasslands." In *Physiological Ecology of North American Plant Communities*, edited by Brian F. Chabot and Harold A. Mooney. New York: Chapman and Hall, 1985.

Rosenberg, Norman J. "Climate of the Great Plains Region of the United States." *Great Plains Quarterly* 7 (Winter 1986): 22–32.

Sala, O.E., and W.K. Lauenroth. "Small Rainfall Events: An Ecological Role in Semiarid Regions." *Oecologia* 53 (1982): 301–4.

Sandburg, Carl. *Cornhuskers*. New York: Henry Holt, 1918.

Sosebee, Ronald E., et al., eds. *Rangeland Plant Physiology*. Denver: Society for Range Management, 1977.

Trenberth, Kevin E., et al. "Origins of the 1988 North American Drought." *Science* 242 (1998): 1640–45.

———, and Christian J. Guillemot. "Physical Processes Involved in the 1988 Drought and 1993 Floods in North America." *Journal of Climate* 9 (1996): 1288–98.

U.S. Global Change Research Program. "US National Assessment of the Potential Consequences of Climate Variability and Change, Mega Region: Great Plains." www.usgcrp.gov/usgcrp/nacc/greatplains.htm/.

Woodhouse, Connie A., and Jonathan T. Overpeck. " 2000 Years of Drought Variability in the Central United States." *Bulletin of the American Meteorological Society* 70 (1988): 2693–714.

CHAPTER 4: SECRETS OF THE SOIL

Blair, J.M., et al. "Responses of Grassland Soil Invertebrates to Natural and Anthropogenic Disturbances." In *Invertebrates as Webmasters in Ecosystems*, edited by D.C. Coleman and P.F. Hendrix, 43–71. New York: CABI Publishers, 2000.

Buol, S.W., et al. *Soil Genesis and Classification*. 5th ed. Ames: Iowa State University Press, 2003.

Callahan, M.A., Jr., and J.M. Blair. "Influence of Differing Land Management on the Invasion of North American Tall Grass Prairie Soils by European Earthworms." *Pedobiologia* 43 (1999): 507–12.

Capinera, John L., ed. *Integrated Pest Management on Rangeland: A Shortgrass Prairie Perspective*. Boulder: Westview, 1987.

Gregg, Robert E. *The Ants of Colorado*. Boulder: University of Colorado Press, 1963.

Heat-Moon, William Least. *PrairyErth: A Deep Map*. Boston: Houghton Mifflin, 1991.

James, Samuel W. "Soil, Nitrogen, Phosphorus, and Organic Matter Processing by Earthworms in Tallgrass Prairie." *Ecology* 72 (1991): 2101–9.

———, and T.R. Seastedt. "Nitrogen Mineralization by Native and Introduced Earthworms: Effects on Big Bluestem Growth." *Ecology* 6 (1986): 1094–97.

Kohnke, Helmut, and D.P. Franzmeier. *Soil Science Simplified*. Prospect Heights: Waveland, 1995.

Paul, E.A., et al. "Micro-organisms." In *Grassland Ecosystems of the World: Analysis of Grasslands and Their Uses*, edited by R.T. Coupland, 87–96. Cambridge: Cambridge University Press, 1979.

Pavlychenko, T.K. *Root Systems of Certain Forage Crops in Relation to the Management of Agricultural Soils*. Ottawa: National Research Council, Canada, No. 1088, 1942.

Reynolds, John W. "Status of Exotic Earthworm Systematics and Biogeography in North America." In *Earthworm Ecology and Biogeography in North America*, edited by P.F. Hendrix, 1–27. Boca Raton: Lewis, 1995.

Rice, Elroy L. "Allelopathy and Grassland Improvement." *The Grasses and Grasslands of Oklahoma*, 90–111. Annals of the Oklahoma Academy of Science, No. 6, 1974.

Stanton, N.L. "The Underground in Grasslands." *Annual Review of Ecology and Systematics* 19 (1988): 573–89.

Trager, James C. "An Introduction to Ants (Formicidae) of the Tallgrass Prairie." *Missouri Prairie Journal* 18 (1998): 4–8. www.npwrc.usgs.gov/resource/insects/ants/index.htm

Van der Heijden, Marcel G.A., et al. "The Unseen Majority: Soil Microbes as Drivers of Plant Diversity and Productivity in Terrestrial Ecosystems." *Ecology Letters* 11 (2008): 296–310.

Wheeler, George C., and Jeannette Wheeler. *The Ants of North Dakota*. Grand Forks: University of North Dakota, 1971....

Wolfe, David W. *Tales From the Underground: A Natural History of Subterranean Life*. Cambridge: Perseus, 2001.

CHAPTER 5: HOME ON THE RANGE

Allen, Steve. "Coyotes on the Move: Changes in Distribution of Coyotes and Red Fox in North Dakota." *North Dakota Outdoors* 58 (1996): 6–11. www.npwrc.usgs.gov/resource/mammals/coyotes/index.htm

Brown, James H., and Edward J. Heske. "Control of a Desert-Grassland Transition by a Keystone Rodent Guild." *Science* 250 (1990) 1705–7.

Byers, John A. *American Pronghorn: Social Adaptations and the Ghosts of Predators Past*. Chicago: University of Chicago Press, 1997.

Collins, Alan R., et al. "An Economic Analysis of Black-tailed Prairie Dog Control." *Journal of Range Management* 37 (1984): 358–61.

Copprock, D.L., et al. "Plant-Herbivore Interactions in a North American Mixed-Grass Prairie. II. Responses of Bison to Modification of Vegetation by Prairie Dogs." *Oecologia* 56 (1983): 10–15.

Derner, Justin D., et al. "Are Livestock Weight Gains Affected by Black-Tailed Prairie Dogs?" *Frontiers in Ecology and the Environment* 4 (2006): 459–64.

Dobkin, David S. *Conservation and Management of Neotropical Migrant Landbirds in the Northern Rockies and Great Plains*. Moscow: University of Idaho Press, 1994.

Fleischner, Thomas L. "Ecological Costs of Livestock Grazing in Western North America." *Conservation Biology* 8 (1994): 629–44.

Frank, Douglas A., et al. "The Ecology of the Earth's Grazing Ecosystems." *BioScience* 48 (1998): 513–21.

———, and Samuel J. McNaughton. "Evidence for the Promotion of Aboveground Grassland Production by Native Large Herbivores in Yellowstone National Park." *Oecologia* 96 (1993): 157–61.

Gillihan, S.W., et al. *Sharing Your Land with Shortgrass Prairie Birds*. Brighton: Rocky Mountain Bird Observatory, 2001.

Goriup, Paul D. *Ecology and Conservation of Grassland Birds*. Cambridge: International Council for Bird Preservation, 1988.

Hanski, Ilkka, and Yves Cambefort. *Dung Beetle Ecology*. Princeton: Princeton University Press, 1991.

Higley, Brewster. "My Western Home." *Kirwin [Kansas] Chief* 3 (February 26, 1876), n.p.

Holechek, Jerry L., et al. *Range Management: Principles and Practices*. 5th ed. Upper Saddle River: Prentice Hall, 2004.

Hoogland, John L. *The Black-Tailed Prairie Dog: Social Life of a Burrowing Mammal*. Chicago: University of Chicago Press, 1995.

Kamler, Jan F., and Warren B. Ballard. "A Review of Native and Nonnative Red Foxes in North America." *Wildlife Society Bulletin* 30 (2002): 270–79.

Knapp, Alan K., et al., eds. *Grassland Dynamics: Long-Term Ecological Research in Tallgrass Prairie*. New York: Oxford University Press, 1998.

Knopf, Fritz L. "Changing Landscapes and the Cosmopolitism of the Eastern Colorado Avifauna." *Wildlife Society Bulletin* 14 (1986): 132–42.

Koford, Carl B. "Prairie Dogs, Whitefaces, and Blue Grama." *Wildlife Monographs* 3 (1958): 1–80.

Johnson, Douglas H., and Alan B. Sargeant. 1977. "Impact of Red Fox Predation on the Sex Ratio of Prairie Mallards." U.S. Fish and Wildlife Service Wildlife Research Report 6. www.npwrc.usgs.gov/resource/mammals/sexrationsexratio.htm.

Lockwood, Jeffrey A., and Larry D. DeBrey. "A Solution for the Sudden and Unexplained Extinction of the Rocky Mountain Grasshopper (Orthoptera: Acrididae)." *Environmental Entomology* 19 (1990): 1194–1205.

Marriot, Alice, and Carol K. Rachlin. *Plains Indian Mythology*. New York: New American Library, 1985.

Michener, Gail R., and Josef K. Schmutz. "Richardson's Ground Squirrel (also known as gophers." http://research.uleth.ca/rgs/

Minta, Steven E., et al. "Hunting Associations Between Badgers (*Taxidea taxus*) and Coyotes (*Canis latrans*)." *Journal of Mammalogy* 73 (1992): 814–20.

Polley, H. Wayne, and Scott L. Collins. "Relationships of

Vegetation and Environment in Buffalo Wallows."
American Midland Naturalist 112 (1984): 178–86.

Saskatchewan Wetland Conservation Corporation.
A Land Manager's Guide to Grassland Birds of Saskatchewan.
Regina: SWCC, 2002.

Savage, Candace. "Return of the Ferret."
Canadian Geographic 129 (Dec. 2009): 58–70.

Tannas, Kathy. *Common Plants of the Western Rangelands.* 2
vols. Lethbridge: Lethbridge Community
College, 1997.

U.S. Fish and Wildlife Service. "Twelve-Month
Finding on a Petition to List the Black-tailed Prairie
Dogs as Threatened or Endangered, 12/03/2009." www.
fws.gov/endangered/

Vavra, Martin, et al. *Ecological Implications of Livestock
Herbivory in the West.* Denver: Society for Range
Management, 1994.

Vickery, Peter D., and James R. Herkert, eds. *Ecology and
Conservation of Grassland Birds of the Western Hemisphere.*
Camarilla: Cooper Ornithological Society, 1999.

Wagner, Frederic H. "Half Century of American Range
Ecology and Management: a Retrospective." In
*Foundations of Environmental Sustainability: the Coevolution
of Science and Policy,* 129–47. Edited by
Larry L. Rockwood, et al. Oxford: Oxford University
Press, 2008.

CHAPTER 6: WATER OF LIFE

Anderson, Michael G., and Lisa G. Sorenson. "Global
Climate Change and Waterfowl: Adaptation in the Face
of Uncertainty." *North American Wildlife and Natural
Resources Conference* 66 (2001): 300–19.

Batzer, Donald P., et al., eds. *Invertebrates in Freshwater
Wetlands of North America: Ecology and Management.* New
York: John Wiley and Sons, 1999.

Beauchamp, Wendy D., et al. "Long Term Declines in Nest
Success of Prairie Ducks." *Journal of Wildlife Management*
60 (1996): 247–57.

Bent, Arthur C. *Life Histories of Familiar North American Birds.*
http://birdsbybent.com.

Bolen, Eric G., et al. "Playa Lakes: Prairie Wetlands of the
Southern High Plains." *BioScience* 39 (1989): 615–23.

Carpenter, David. *Fishing in the West: A Guide to Alberta,
Saskatchewan and Manitoba.* Saskatoon: Western
Producer Prairie Books, 1984.

Cross, Frank B., and Randall E. Moss. "Historic Changes
in Fish Communities and Aquatic Habitats in Plains
Streams of Kansas." In *Community and Evolutionary
Ecology of North American Stream Fishes,* edited by
William J. Matthews and David C. Heins, 155–77.
Norman: University of Oklahoma Press, 1987.

Cummings, K.S., and C.A. Mayer. *Freshwater Mussels of
the Midwest.* Illinois Natural History Society Survey
Manual 5, 1992. www.inhs.illinois.edu/animals_plants/
mollusk/fieldguide.

Dahl, Thomas E. *Status and Trends of Wetlands in the
Conterminous United States 1986 to 1997.* Washington,
D.C.: U.S. Department of the Interior, Fish and Wildlife
Service, 2000.

Ferril, Thomas Hornsby. *New and Selected Poems.* New York:
Harper, 1952.

Fischer, Jesse R., and Craig P. Paukert. "Historical and
Current Environmental Influences on an Endemic
Great Plains Fish." *American Midland Naturalist* 159
(2008): 364–77.

Galatowitsch, Susan M., and Arnold G. Van der Valk.
Restoring Prairie Wetlands: An Ecological Approach. Ames:
Iowa State University Press, 1994.

Gilbert, Lee, et al. *Atlas of North American Freshwater Fishes.*
Raleigh: North Carolina State Museum of Natural
History, 1980.

Johnsgard, Paul A. *Crane Music: A Natural History of American
Cranes.* Washington, D.C.: Smithsonian, 1991.

———. *The Platte: Channels in Time.* Lincoln: University of
Nebraska Press, 1984.

Johnson, Barry L., et al. "Past, Present, and Future
Concepts in Large River Ecology." *BioScience* 45 (1995):
134–41.

McCafferty, W. Patrick. *Aquatic Entomology: The Fishermen's
and Ecologists' Illustrated Guide to Insects and Their Relatives.*
Boston: Science Books International, 1981.

Murkin, Henry R. *Prairie Wetland Ecology: The Contribution
of the Marsh Ecology Research Program.* Ames: Iowa State
University Press, 2000.

Pallid Sturgeon Recovery Coordinator. "Pallid Sturgeon (*Scaphirynchus albus*) 5-Year Review: Summary and Evaluation." Billings, Montana, June, 2007.

Petrik, Paula. "Remaking the World: A Sioux Creation Story, ca. 1910." www.archiva.net/hist120ay02/ readings/ lakotacreation.

Poiani, Karen A., and W. Carter Johnson. "Global Warming and Prairie Wetlands." *BioScience* 41 (1991): 613–18.

Power, Greg. "The Missouri River System's 'Other' Fish." *North Dakota Outdoors* 56 (1993): 2–9. www.npwrc.usgs. gov/resource/fish/othrfish/index.htm.

Reynolds, Ronald E., et al. "Impact of the Conservation Reserve Program on Duck Recruitment in the U.S. Prairie Pothole Region." *Journal of Wildlife Management* 65 (2001): 765–80.

Richkus, Kenneth D. *Northern Pintail Nest Site Selection, Nest Success, Renesting Ecology, and Survival in the Intensively Farmed Prairies of Southern Saskatchewan: An Evaluation of the Ecological Trap Hypothesis.* Ph.D. diss., Louisiana State University, 2002.

Ryckman, Fred. "The Confluence." *North Dakota Outdoors* 8 (March 2000): 8–10.

Seburn, Carolyn N.L., et al. "Northern Leopard Frog Dispersal in Relation to Habitat." *Amphibians in Decline: Canadian Studies of a Global Problem. Herpetological Conservation* 1 (1997): 64–72.

Spencer, John R., and Nils Moller Andersen. "Biology of Water Striders: Interactions Between Systematics and Ecology." *Annual Review of Entomology* 39 (1994): 101–28.

U.S. Fish and Wildlife Service. *Waterfowl Population Status*, 2009. U.S. Department of the Interior, Washington, D.C. www.fws.gov/migratorybirds/.

Van der Valk, A.G. "The Role of Seed Banks in the Vegetation Dynamics of Prairie Glacial Marshes." *Ecology* 59 (1978): 322–35.

———. "Succession in Wetlands: A Gleasonian Approach." *Ecology* 63 (1981): 688–96.

———, and C.B. Davis. "The Seed Banks of Prairie Glacial Marshes." *Canadian Journal of Botany* 54 (1976): 1832–38.

Water Science and Technology Board. *The Missouri River Ecosystem: Exploring the Prospects for Recovery.* Washington, D.C.: National Academy Press, 2002. http://books.nap.edu/catalog/1027.html.

Weller, Milton W. *Freshwater Marshes: Ecology and Wildlife Management.* Minneapolis: University of Minnesota Press, 1994.

World Wildlife Fund. *Canada's Rivers at Risk: Environmental Flows and Canada's Freshwater Future*, 2009. http://wwf.ca/conservation/freshwater/riversatrisk. cfm.

CHAPTER 7: PRAIRIE WOODLANDS

Anderson, C., and D.W. McShea. "Intermediate-level Parts in Insect Societies: Adaptive Structures That Ants Build Away from the Nest." *Insectes Sociaux* 48 (2001): 291–301.

Archer, Steve. "Have Southern Texas Savannas Been Converted to Woodlands in Recent History?" *American Naturalist* 134 (1989): 544–61.

Bird, Ralph D. *Ecology of the Aspen Parkland of Western Canada in Relation to Land Use.* Ottawa: Research Branch, Canada Department of Agriculture, 1961.

Black Elk. *Black Elk Speaks: Being the Life Story of a Holy Man of the Ogalala Sioux, as Told Through John G. Neihardt.* New York: Morrow, 1932.

Bradley, Cheryl E., and Derald G. Smith. "Plains Cottonwood Recruitment and Survival on a Prairie Meandering River Floodplain, Milk River, Southern Alberta and Northern Montana." *Canadian Journal of Botany* 64 (1989): 1433–42.

Bragg, Thomas B., and Annehara K. Tatschi. "Changes in Flood-Plain Vegetation and Land Use Along the Missouri River from 1826 to 1972." *Environmental Management* 1 (1977): 343–48.

Cable, Ted T. "Windbreaks, Wildlife, and Hunters." In *Wildlife and Habitats in Managed Landscapes.* Washington, D.C.: Island Press, 1991.

Campbell, B.D., et al. "A Synthesis of Recent Global Change Research on Pasture and Rangeland Production: Reduced Uncertainties and Their Management Implications." *Agriculture Ecosystems and Environment* 82 (2000): 39–55.

Droze, Wilmon H. *Trees, Prairies, and People: A History of Tree Planting in the Plains States*. Denton: Texas Woman's University, 1977.

Elliott, Phillip F. "Cowbird Parasitism in the Kansas Tallgrass Prairie." *Auk* 95 (1978): 161–67.

Fitch, L., and B.W. Adams. "Can Cows and Fish Co-Exist?" *Canadian Journal of Plant Science* 78 (1998): 191–98.

Fox, L. B., et al. *Habitat Guidelines for Mule Deer: Great Plains Ecoregion*. Mule Deer Working Group, Western Association of Fish and Wildlife Agencies, 2009.

Houston, C. Stuart. "The Spread of the Western Kingbird Across the Prairies." *Blue Jay* 37 (1979): 149–57.

———, and Mary I. Houston. "Saskatchewan Bird Species Which Increased with Settlement." *Blue Jay* 55 (1997): 90–97.

Johnsgard, Paul A. "Cranes of the World in 2008: a Supplement to *Crane Music*." *Papers in Ornithology*, University of Nebraska-Lincoln, 2008. http//:digitalcommons.unl.edu/biosciornithology/45.

Johnson, W. Carter. "Woodland Expansion in the Platte River, Nebraska: Patterns and Causes." *Ecological Monographs* 64 (1994): 45–84.

Kauffman, J. Boone, and W.C. Krueger. "Livestock Impacts on Riparian Ecosystems and Streamside Management Implications: A Review." *Journal of Range Management* 37 (1984): 430–38.

Kaul, Robert B., et al. "The Niobrara River Valley: A Postglacial Migration Corridor and Refugium of Forest Plans and Animals in the Grasslands of Central North America." *Botanical Review* 54 (1998): 44–75.

Reily, Peggy W., and W. Carter Johnson. "The Effects of Altered Hydrologic Regime on Tree Growth Along the Missouri River in North Dakota." *Canadian Journal of Botany* 60 (1982): 2410–23.

Ripple, William J., and Robert L. Beschta. "Hardwood Tree Decline Following Large Carnivore Loss on the Great Plains, USA." *Frontiers in Ecology and the Environment* 5 (2007): 241–46.

Rood, Stewart B., and John M. Mahoney. "Collapse of Riparian Poplar Forests Downstream Dams in Western Prairies: Probable Causes and Prospects for Mitigation." *Environmental Management* 14 (1990): 451–64.

Streng, D.R., and P.A. Harcombe. "Why Don't East Texas Savannas Grow Up to Forest?" *American Midland Naturalist* 108 (1982): 287–94.

Svejcar, Tony. "Riparian Zones: 1) What Are They and How Do They Work?" *Rangelands* 19 (1997): 4–7.

———. "Riparian Zones: 2) History and Human Impacts." *Rangelands* 19 (1997): 8–12.

Webster, R.P., and M.C. Nielsen. "Myrmecophily in the Edward's Hairstreak Butterfly *Sattyrium edwardsii* (Lyncaenidae)." *Journal of the Lepidopterists' Society* 38 (1984): 124–33.

Wells, Philip V. "Scarp Woodlands, Transported Grassland Soils, and Concept of Grassland Climate in the Great Plains Region." *Science* 148 (1965): 246–49.

Wilson, Roger E. "Succession in Stands of *Populus deltoides* Along the Missouri River in Southeastern South Dakota." *American Midland Naturalist* 83 (1970): 331–42.

Winter, Maiken. "Nesting Biology of Dickcissels and Henslow's Sparrows in Southwestern Missouri Prairie Fragments." *Wilson Bulletin* 111 (1999): 515–27.

———, and John Faaborg. "Patterns of Area Sensitivity in Grassland-Nesting Birds." *Conservation Biology* 13: 1424–36.

CHAPTER 8: THE NATURE OF FARMING

Altieri, Miguel A. "How Best Can We Use Biodiversity in Agroecosystems?" *Outlook on Agriculture* 20 (2001): 15–23.

Baker, Herbert G. "The Evolution of Weeds." *Annual Review of Ecology and Systematics* 5 (1974): 1–24.

Berry, Wendell. "Hope." In *Fatal Harvest: The Tragedy of Industrial Agriculture*, edited by Andrew Kimbrell, 373–75. Washington, D.C.: Island Press, 2002.

Best, Louis B. "Conservation Tillage: Ecological Traps for Nesting Birds?" *Wildlife Society Bulletin* 14 (1986): 308–17.

———, Timothy Bergin, and Kathryn E. Freemark. "Influence of Landscape Composition on Bird Use of Rowcrop Fields." *Journal of Wildlife Management* 65 (2001): 442–49.

———, Kathryn Freemark, et al. "A Review and Synthesis of Habitat Use by Breeding Birds in Agricultural Landscapes of Iowa." *American Midland Naturalist* 134 (1995): 1–29.

———, Robert C. Whitmore, and Gary M. Booth. "Use of Cornfields by Birds During the Breeding Season: The Importance of Edge Habitat." *American Midland Naturalist* 123 (1990): 84–99.

Bryan, Georgia G., and Louis B. Best. "Bird Abundance and Species Richness in Grassed Waterways in Iowa Rowcrop Fields." *American Midland Naturalist* 126 (1991): 90–102.

Buchmann, Stephen L., and Gary Paul Nabham. *The Forgotten Pollinators*. Washington, D.C.: Island Press, 1996.

Calpas, James, and Dan Johnson. "Grasshopper Management." *Agri-Facts*. Alberta Agriculture, Food and Rural Development, March, 2003.

Camp, Martha, and Louis B. Best. "Bird Abundance and Nesting in CRP Fields and Cropland in the Midwest: A Regional Approach." *Wildlife Society Bulletin* 25 (1997): 864–77.

Carcamo, Hector A., et al. "Farming and Ground Beetles: Effects of Agronomic Practice on Populations and Community Structure." *Canadian Entomologist* 127 (1995): 123–40.

Clark, Robert G. et al. "Making a Living on the Edge: Beneficial Effects of Field Boundary Habitats for Biodiversity and Agriculture." *Canadian Entomologist* 130 (1998): 321–36.

Cox, Robert R., Jr. "Spring-Staging Ecology of Waterfowl in Nebraska—Then Versus Now." *Waterfowl* 12 (2000): 18–19.

Curry, Philip S. *Bumble Bees of Saskatchewan (Hymenoptera: Apidae)*. Natural History Contributions No. 5, Saskatchewan Museum of Natural History, 1984.

Erwin, Terry L., et al. *Carabid Beetles: Their Evolution, Natural History, and Classification*. London: Dr. W. Junk BV Publishers, 1979.

Jackson, Dana L., and Laura L. Jackson, eds. *The Farm as Natural Habitat: Reconnecting Food Systems with Ecosystems*. Washington, D.C.: Island Press, 2002.

Jackson, Wes: "Natural Systems Agriculture: A Truly Radical Alternative." *Agriculture Ecosystems and Environment* 88 (2002): 111–17.

———. *New Roots for Agriculture*. San Francisco: Friends of the Earth, 1980.

Johnson, Dan. "Band-winged Grasshoppers of the Canadian Prairies and Northern Great Plains." *Arthropods of Canadian Grasslands* 7 (2001): 5–12. http://people.uleth.ca/~dan.johnson/htm/dj_gh_guide.htm.

———. "Spur-Throated Grasshoppers of the Canadian Prairies and Northern Great Plains." *Arthropods of Canadian Grasslands* 8 (2002): 16–25. http://people.uleth.ca/~dan.johnson/htm/dj_gh_guide.htm.

———. "Slant-faced Grasshoppers of the Canadian Prairies and Northern Great Plains." *Arthropods of Canadian Grasslands* 9 (2003): 5–16. http://people.uleth.ca/~dan.johnson/htm/dj_gh_guide.htm.

Johnson, Douglas H., and Lawrence D. Igl. "Contributions of the Conservation Reserve Program to Populations of Breeding Birds in North Dakota." *Wilson Bulletin* 107 (1995): 709–18.

Kirk, David A., et al. "Past and Current Attempts to Evaluate the Role of Birds as Predators of Insect Pests in Temperate Agriculture." *Current Ornithology* 13 (1996): 175–263.

Kirk, Vernon M. "Ground Beetles in Cropland in South Dakota." *Annals of the Entomological Society of America* 64 (1971): 238–41.

Matson, P.A., et al. "Agricultural Intensification and Ecosystem Properties." *Science* 277 (1997): 504–9.

McCoy, Timothy D., et al. "Conservation Reserve Program: Source or Sink Habitat for Grassland Birds in Missouri?" *Journal of Wildlife Management* 63 (1999): 530–38.

McMaster, D. Glen, and Stephen K. Davis. "An Evaluation of Canada's Permanent Cover Program: Habitat for Grassland Birds?" *Journal of Field Ornithology* 723 (2001): 195–210.

Owens, R.A., and M.T. Myres. "Effects of Agriculture Upon Populations of Native Passerine Birds of an Alberta Fescue Grassland." *Canadian Journal of Zoology* 51 (1971): 697–713.

Rodenhouse, Nicholas L., and Louis B. Best. "Breeding Ecology of Vesper Sparrows in Corn and Soybean Fields." *American Midland Naturalist* 110 (1983): 265–75.

Samson, F.B., and F.L. Knopf. "Prairie Conservation in North America." *BioScience* 44 (1994): 418–21.

Shutler, Dave, et al. "Bird Communities of Prairie Uplands and Wetlands in Relation to Farming Practices in Saskatchewan." *Conservation Biology* 14 (2000): 1441–51.

Soule, Judith D., and Jon K. Piper. *Farming in Nature's Image: An Ecological Approach to Agriculture.* Washington, D.C.: Island Press, 1992.

Svedarsky, W. Daniel, et al., eds. *The Great Prairie Chicken: A National Look.* University of Minnesota Miscellaneous Publication 99-1999, 1999.

Tilman, David. "The Greening of the Green Revolution." *Nature* 396 (1998): 211–12.

Welsh R. *The Economics of Organic Grain and Soybean Production in the Midwestern United States.* Henry A. Wallace Institute Policy Studies Report No. 13, May 1999.

Wooley, James B., Jr., et al. "Impacts of No-Till Row Cropping on Upland Wildlife." *North American Wildlife and Nature Resources Conference* 50 (1985): 157–66.

CHAPTER 9: LONG-RANGE FORECAST

Askins, Robert A., et al. "Conservation of Grassland Birds in North America: Understanding Ecological Processes in Different Regions." White Paper for AOU Conservation Committee, 2007. www.aou.org/committees/docs/ConservationAddn4.pdf.

Bagley, Catherine, et al. "Organic Agriculture and the Global Food Supply." *Renewable Agriculture and Food Systems* 22 (2006): 86–108.

Black Elk. *Black Elk Speaks: Being the Life Story of a Holy Man of the Ogalala Sioux, as Told Through John G. Neihardt.* New York: Morrow, 1932.

Blancher, Peter. *Importance of North America's Grasslands to Birds.* Commission for Environmental Cooperation, 2003.

Brower, Jennifer. *Lost Tracks: Buffalo National Park, 1909–1939.* Edmonton: Athabasca University Press, 2008.

Conner, Richard, et al. *United States Grasslands and Related Resources: An Economic and Biological Trends Assessment.* Prepared for the National Cattlemen's Beef Association, the Nature Conservancy and Ducks Unlimited, 2001. http://imr.tamu.edu/pdf/grasslands_high.pdf.

Freese, Curtis, et al. *New Directions for the Prairie Economy: Connecting Conservation and Rural Development in the Northern Great Plains.* World Wildlife Fund, 2009. www.worldwildlife.org/what/wherewework/ngp/publications.html.

Gates, C. Cormack, et al. American Bison: Status Survey and Conservation Guidelines 2010, International Union for Conservation of Nature, Species Survival Commission. http://cmsdata.iucn.org/downloads/american_bison_report.pdf.

I-See-O, as quoted by Dan Flores, "Sky, Mesa and the Illano Estacado Bioregion." www-personal.ksu.edu/~jsherow/flores.htm.

Kimbrell, Andrew, ed. *Fatal Harvest: The Tragedy of Industrial Agriculture.* Washington, D.C.: Island Press, 2002.

Knight, Richard L., et al. *Ranching West of the 100th Meridian: Culture, Ecology, and Economics.* Washington, D.C.: Island Press, 2002.

Knopf, Fritz L., and Fred B. Samson, eds. *Ecology and Conservation of Great Plains Vertebrates.* New York: Springer, 1996.

Martin, Leanne M., et al. "An Assessment of Grassland Restoration Success Using Species Diversity Components." Journal of Applied Ecology 42 (2005): 327–36. McCracken, Jon D. "Where the Bobolinks Roam: The Plight of North America's Grassland Birds." *Biodiversity* 6 (2005): 20–29. www.bsc-eoc.org/download/GrasslandBirdsMcCracken.pdf.

Millet, Bruce, et al. "Climate trends of the North American Prairie Pothole Region 1906-2000." *Climatic Change* 93 (2009): 243–47.

Moehrenschlager, Axel, and Cynthis Moehrenschlager. *Population Census of Reintroduced Swift Foxes in Canada and Northern Montana 2005/2006.* Calgary: Centre for Conservation Research, Calgary Zoo, 2006.

Northern Plains Conservation Network. *Oceans of Grass: A Conservation Assessment for the Northern Great Plains,* 2004. www.worldwildlife.org/what/wherewework/ngp/publications.html.

Ostlie, W.R., et al. *The Status of Biodiversity in the Great Plains*. The Nature Conservancy, Arlington: 1997.

Packard, Stephen, and Cornelia F. Mutel. *The Tallgrass Restoration Handbook: For Prairies, Savannas and Woodlands*. Washington, D.C.: Island Press, 1997.

Power, Thomas Michael, and Richard Barrett. *Post-Cowboy Economics: Pay and Prosperity in the New American West*. Washington, D.C.: Island Press, 2001.

Ries, Leslie, et al. "Conservation Value of Roadside Prairie Restoration to Butterfly Communities." *Conservation Biology* 15 (2001): 401–11.

Samson, Fred B., and Fritz L. Knopf, eds. *Prairie Conservation: Preserving North America's Most Endangered Ecosystem*. Washington, D.C.: Island Press, 1996.

Sayre, Nathan F. *Working Wilderness: The Malpai Borderlands Group and the Future of the Western Range*. Tucson: Rio Nuevo, 2005.

Shirley, Shirley. *Restoring the Tallgrass Prairie*. Iowa City: University of Iowa Press, 1994.

Smith, Elwin G., et al. "Profitability and Risk of Organic Production Systems in the Northern Great Plains." *Renewable Agriculture and Food Systems* 19 (2004): 152–58.

U.S. Global Change Research Program. "U.S. National Assessment of the Potential Consequences of Climate Variability and Change, Mega Region: Great Plains." www.usgcrp.gov/usgcrp/nacc/greatplains.htm/.

{ APPENDICES }

Scientific Names

Common names for plants and animals often vary
from place to place. To help prevent confusion, this list
provides the currently accepted scientific names for
extant species found on the Geat Plains. Where common
names refer to subspecies, scientific names have been
provided to the species level only. Because taxonomy
is an ever-evolving science, the names given here may
differ somewhat from those found in older field guides
and other reference books.

American avocet, *Recurvirostra americana*
American badger, *Taxidea taxus*
American black bear, *Ursus americanus*
American burying beetle, *Nicrophorus americanus*
American crow, *Corvus brachyrhynchos*
American elm, *Ulmus americana*
American goldfinch, *Carduelis tristis*
American kestrel, *Falco sparverius*
American robin, *Turdus migratorius*
American wigeon, *Anas americana*
Arkansas River shiner, *Notropis girardi*
armadillo, *Dasypus novemcinctus*
Ashe juniper, *Juniperus ashei*
ash-throated flycatcher, *Myiarchus cinerascens*
aspen poplar, *Populus tremuloides*
awned wheatgrass, *Elymus trachycaulus*

Baird's sandpiper, *Calidris bairdii*
Baird's sparrow, *Ammodramus bairdii*
balsam poplar, *Populus balsamifera*
Baltimore oriole, *Icterus galbula*
banded argiope, *Argiope trifasciata*
barn swallow, *Hirundo rustica*

barnyard grass, *Echinochloa crus-galli*
beaver, *Castor canadensis*
big bluestem, *Andropogon gerardii*
bighorn sheep, *Ovis canadensis*
bison, *Bison bison*
black-bellied plover, *Pluvialis squatarola*
black-billed magpie, *Pica hudsonia*
black bullhead, *Ameiurus melas*
black-capped chickadee, *Poecile atricapillus*
black crappie, *Pomoxis nigromaculatus*
black-crested titmouse, *Baeolophus atricristatus*
black-footed ferret, *Mustela nigripes*
black-headed grosbeak, *Pheucticus melanocephalus*
Black Hills red-bellied snake, *Storeria occipitomaculata*
blacknose shiner, *Notropis heterolepis*
black-tailed jackrabbit, *Lepus californicus*
black-tailed prairie dog, *Cynomys ludovicianus*
black tern, *Chlidonias niger*
blowout grass, *Redfieldia flexuosa*
blowout penstemon, *Penstemon haydenii*
bluebunch wheatgrass, *Pseudoroegneria spicata*
bluebur, *Lappula echinata*
blue grama, *Bouteloua gracilis*
blue jay, *Cyanocitta cristata*
blue-stem prickly-poppy, *Argemone albiflora*
blue vervain, *Verbena hastata*
blue-winged teal, *Anas discors*
bobcat, *Lynx rufus*
bobolink, *Dolichonyx oryzivorus*
box elder, *Acer negundo*
brassy minnow, *Hybognathus hankinsoni*
broomweed, *Amphiachyris dracunculoides*
brown-eyed susan, *Rudbeckia triloba*
brown-headed cowbird, *Molothrus ater*

brown-spotted range grasshopper, *Psoloessa delicatula*
buffalo grass, *Buchloe dactyloides*
bufflehead, *Bucephala albeola*
Bullock's oriole, *Icterus bullockii*
bur oak, *Quercus macrocarpa*
burrowing owl, *Athene cunicularia*
butterfly milkweed, *Asclepias tuberosa*

California gull, *Larus californicus*
Canada goose, *Branta canadensis*
Canada thistle, *Cirsium arvense*
Canada wild rye, *Elymus canadensis*
canvasback, *Aythya valisineria*
Carlinian snapper grasshopper, *Circotettix carlinianus*
Carolina chickadee, *Poecile carolinensis*
cat's claw mimosa, *Mimosa aculeaticarpa*
chestnut-collared longspur, *Calcarius ornatus*
chinook salmon, *Oncorhynchus tshawytscha*
chipping sparrow, *Spizella passerina*
chokecherry, *Prunus virginiana*
claret cup, *Echinocereus triglochidiatus*
clear-winged grasshopper, *Camnula pellucida*
collared peccary, *Pecari tajacu*
common goldeneye, *Bucephala clangula*
common grackle, *Quiscalus quiscula*
common merganser, *Mergus merganser*
common nighthawk, *Chordeiles minor*
concho water snake, *Nerodia paucimaculata*
coontail, *Ceratophyllum demersum*
cougar, *Puma concolor*
coyote, *Canis latrans*
creosote bush, *Larrea tridentata*
crested wheatgrass, *Agropyron cristatum*
Culver's root, *Veronicastrum virginicum*
curly dock, *Rumex crispus*

deer mouse, *Peromyscus maniculatus*
dickcissel, *Spiza americana*
Douglas-fir, *Pseudotsuga menziesii*
downy woodpecker, *Picoides pubescens*
dunes sagebrush lizard, *Sceloporus arenicolus*

eared grebe, *Podiceps nigricollis*
eastern bluebird, *Sialia sialis*
eastern kingbird, *Tyrannus tyrannus*
eastern meadowlark, *Sturnella magna*
eastern red bat, *Lasiurus borealis*
eastern red cedar, *Juniperus virginiana*
eastern screech-owl, *Megascops asio*
eastern spotted skunk, *Spilogale putorius*
eastern towhee, *Pipilo erythrophthalmus*
eastern wood-pewee, *Contopus virens*
eastern woodrat, *Neotoma floridana*
Edwards' hairstreak butterfly, *Satyrium edwardsii*
Engelmann spruce, *Picea engelmannii*
English grain aphid, *Macrosiphum avenae*
European corn borer, *Ostrinia nubilalis*
European starling, *Sturnus vulgaris*

ferruginous hawk, *Buteo regalis*
flathead chub, *Platygobio gracilis*
flat-topped aster, *Doellingeria umbellata*
fox squirrel, *Sciurus niger*
Franklin's ground squirrel,
     Spermophilus franklinii
fringed sagewort, *Artemisia frigida*

gadwall, *Anas strepera*
galleta, *Pleuraphis jamesii*
giant ragweed, *Ambrosia trifida*
golden bean, *Thermopsis rhombifolia*
golden-cheeked warbler, *Dendroica chrysoparia*
golden eagle, *Aquila chrysaetos*
golden-fronted woodpecker, *Melanerpes aurifrons*
grasshopper sparrow, *Ammodramus savannarum*
gray catbird, *Dumetella carolinensis*
gray fox, *Urocyon cinereoargenteus*
great blue heron, *Ardea herodias*
great crested flycatcher, *Myiarchus crinitus*
greater prairie chicken, *Tympanuchus cupido*
greater sage grouse, *Centrocercus urophasianus*
greater short-horned lizard, *Phrynosoma hernandesi*
great horned owl, *Bubo virginianus*
Great Plains toad, *Bufo cognatus*

green ash, *Fraxinus pennsylvanica*
green needlegrass, *Nassella viridula*
green sunfish, *Lepomis cyanellus*
green-winged teal, *Anas crecca*
grizzly bear, *Ursus arctos*
gumbo evening primrose, *Oenothera caespitosa*

harebell, *Campanula rotundifolia*
Harris' checkerspot butterfly, *Chlosyne harrisii*
Henslow's sparrow, *Ammodramus henslowii*
Hessian fly, *Mayetiola destructor*
honey bee, *Apis mellifera*
honey mesquite, *Prosopis glandulosa*
horned lark, *Eremophila alpestris*
hornyhead chub, *Nocomis biguttatus*
house sparrow, *Passer domesticus*
house wren, *Troglodytes aedon*

Indian grass, *Sorghastrum nutans*
Indian paintbrush, *Castilleja indivisa*
indigo bunting, *Passerina cyanea*

javelina, *Pecari tajacu*
johnnydarter, *Etheostoma nigrum*
jumping cactus, *Opuntia fragilis*
June grass, *Koeleria macrantha*

Kentucky blue grass, *Poa pratensis*
kestrel. *See* American kestrel
killdeer, *Charadrius vociferus*

lake sturgeon, *Acipenser fulvescens*
lance-leafed psoralea (scurfpea), *Psoralidium lanceolatum*
largemouth bass, *Micropterus salmoides*
lark bunting, *Calamospiza melanocorys*
lazuli bunting, *Passerina amoena*
leadplant, *Amorpha canescens*
leafy spurge, *Euphorbia esula*
leafy spurge hawkmoth, *Hyles euphorbiae*
LeConte's sparrow, *Ammodramus leconteii*
lesser migratory grasshopper, *Melanoplus sanguinipes*
lesser prairie chicken, *Tympanuchus pallidicinctus*
lesser scaup, *Aythya affinis*

lesser snow goose, *Chen caerulescens*
little bluestem, *Schizachyrium scoparium*
long-billed curlew, *Numenius americanus*

mallard, *Anas platyrhynchos*
Manitoba maple, *Acer negundo*
marbled godwit, *Limosa fedoa*
marsh wren, *Cistothorus palustris*
McCown's longspur, *Calcarius mccownii*
merlin, *Falco columbarius*
Mexican ground squirrel, *Spermophilus mexicanus*
Mississippi kite, *Ictinia mississippiensis*
Missouri milk vetch, *Astragalus missouriensis*
monarch butterfly, *Danaus plexippus*
moose, *Alces alces*
Mormon cricket, *Anabrus simplex*
mound hedgehog-cactus, *Echinocereus triglochidiatus*
mountain bluebird, *Sialia currucoides*
mountain plover, *Charadrius montanus*
mourning dove, *Zenaida macroura*
mule deer, *Odocoileus hemionus*
muskrat, *Ondatra zibethicus*

narrowleaf cottonwood, *Populus angustifolia*
narrow-leaved puccoon, *Lithospermum incisum*
needle-and-thread grass, *Stipa comata*
Nelson's sharp-tailed sparrow, *Ammodramus nelsoni*
neosho madtom, *Noturus placidus*
nighthawk. *See* common nighthawk
nodding beggar-ticks, *Bidens cernua*
nodding onion, *Allium cernuum*
northern cardinal, *Cardinalis cardinalis*
northern flicker, *Colaptes auratus*
northern grasshopper mouse, *Onychomys leucogaster*
northern leopard frog, *Rana pipiens*
northern pike, *Esox lucius*
northern pintail, *Anas acuta*
northern raccoon, *Procyon lotor*
northern shoveler, *Anas clypeata*
northwest poplar, *Populus deltoides* x *Populus balsamifera*
Nuttall's blister beetle, *Lytta nuttalli*
Nuttall's cottontail, *Sylvilagus nuttallii*

olive-backed pocket mouse, *Perognathus fasciatus*
Ord's kangaroo rat, *Dipodomys ordii*
ornate box turtle, *Terrapene ornata*
otter, *Lutra canadensis*
ovenbird, *Seiurus aurocapillus*

paddlefish, *Polyodon spathula*
painted lady butterfly, *Vanessa cardui*
painted turtle, *Chrysemys picta*
pallid sturgeon, *Scaphirhynchus albus*
paper birch, *Betula papyrifera*
partridge pea, *Chamaecrista fasciculata*
pasque flower, *Pulsatilla patens*
pasture sage, *Artemisia frigida*
pasture worm, *Aporrectodea turgida*
pincushion cactus, *Escobaria vivipara*
pink smartweed, *Polygonum amphibium*
piping plover, *Charadrius melodus*
plain pocketbook mussel, *Lampsilis cardium*
plains bison. *See* bison
plains cottonwood, *Populus deltoides*
plains pocket gopher, *Geomys bursarius*
plains prickly pear cactus, *Opuntia polyacantha*
plains spadefoot toad, *Spea bombifrons*
plains wolf. *See* wolf
pocket gopher. *See* plains pocket gopher
ponderosa pine, *Pinus ponderosa*
porcupine, *Erethizon dorsatum*
prairie crocus, *Pulsatilla patens*
prairie dropseed, *Sporobolus heterolepis*
prairie gopher, *Spermophilus richardsonii*
prairie rattlesnake, *Crotalus viridis*
prairie rose, *Rosa arkansana*
prairie sage, *Artemisia ludoviciana*
prairie skink, *Eumeces septentrionalis*
prairie smoke, *Geum triflorum*
prairie sunflower, *Helianthus petiolaris*
prairie three-awn, *Aristida oligantha*
pronghorn, *Antilocapra americana*
purple coneflower, *Echinacea purpurea*
purple prairie clover, *Dalea purpurea*

quackgrass, *Elytrigia repens*
quaking aspen, *Populus tremuloides*

red bat. *See* eastern red bat
red-bellied woodpecker, *Melanerpes carolinus*
red fox, *Vulpes vulpes*
redhead, *Aythya americana*
red-shafted flicker, *Colaptes auratus*
red-tailed hawk, *Buteo jamaicensis*
red-winged blackbird, *Agelaius phoeniceus*
regal fritillary butterfly, *Speyeria idalia*
Richardson's ground squirrel, *Spermophilus richardsonii*
ring-necked duck, *Aythya collaris*
ring-necked pheasant, *Phasianus colchicus*
Rocky Mountain locust, *Melanoplus spretus*
rose-breasted grosbeak, *Pheucticus ludovicianus*
rough fescue, *Festuca hallii, F. scrabella,*
    F. campestris complex
ruddy duck, *Oxyura jamaicensis*
ruffed grouse, *Bonasa umbellus*
Russian olive, *Elaeagnus angustifolia*
Russian thistle, *Salsola tragus*
Russian wheat aphid, *Diuraphis noxia*

sago pondweed, *Potamogeton pectinatus*
salt cedar, *Tamarix ramosissima*
sand bluestem, *Andropogon hallii*
sand dropseed, *Sporobolus cryptandrus*
sand sage, *Artemisia filifolia*
sandhill crane, *Grus canadensis*
sand reedgrass, *Calamovilfa longifolia*
saskatoon, *Amelanchier alnifolia*
Savannah sparrow, *Passerculus sandwichensis*
scarlet butterfly weed, *Gaura coccinea*
scarlet mallow, *Sphaeralcea coccinea*
semipalmated sandpiper, *Calidris pusilla*
serviceberry, *Amelanchier alnifolia*
sharp-tailed grouse, *Tympanuchus phasianellus*
shinnery oak, *Querus havardii*
short-eared owl, *Asio flammeus*
shovelnose sturgeon, *Scaphirhynchus platorynchus*
showy milkweed, *Asclepias speciosa*
sicklefin chub, *Macrhybopsis meeki*

sideoats grama, *Bouteloua curtipendula*
sidewalk beetle, *Pterostichus melanarius*
silverberry, *Elaeagnus commutata*
silverleaf psoralea, *Pediomelum argophyllum*
silver sagebrush, *Artemisia cana*
skeleton weed, *Lygodesmia juncea*
small lupine, *Lupinus pusillus*
smallmouth bass, *Micropterus dolomieu*
Smith's longspur, *Calcarius pictus*
smooth brome, *Bromus inermis*
snowberry, *Symphoricarpos albus*
snow bunting, *Plectrophenax nivalis*
snow goose, *Chen caerulescens*
southern redbelly dace, *Phoxinus erythrogaster*
speckled chub, *Macrhybopsis aestivalis*
spotted ground squirrel, *Spermophilus spilosoma*
spotted sandpiper, *Actitis macularia*
spotted towhee, *Pipilo maculatus*
Sprague's pipit, *Anthus spragueii*
Steller's jay, *Cyanocitta stelleri*
stonecat, *Noturus flavus*
striped skunk, *Mephitis mephitis*
sturgeon chub, *Macrhybopsis gelida*
suckermouth minnow, *Phenacobius mirabilis*
Swainson's hawk, *Buteo swainsoni*
swift fox, *Vulpes velox*
switchgrass, *Panicum virgatum*

tall fescue, *Lolium arundinaceum*
tamarisk, *Tamarix ramosissima*
Texas blind salamander, *Eurycea rathbuni*
Texas bluebonnet, *Lupinus texensis*
thirteen-lined ground squirrel,
     Spermophilus tridecemlineatus
three-flowered avens, *Geum triflorum*
tiger salamander, *Ambystoma tigrinum*
timber milkvetch, *Astragalus miser*
tree swallow, *Tachycineta bicolor*
trembling aspen, *Populus tremuloides*
tufted evening primrose, *Oenothera caespitosa*
tufted titmouse, *Baeolophus bicolor*
Turnbull's grasshopper, *Aeoloplides turnbulli*
two-striped grasshopper, *Melanoplus bivittatus*

upland sandpiper, *Bartramia longicauda*
Uvalde bigtooth maple, *Acer grandidentatum*

vesper sparrow, *Pooecetes gramineus*
Virginia opossum, *Didelphis virginiana*

walleye, *Stizostedion vitreum*
western kingbird, *Tyrannus verticalis*
western meadowlark, *Sturnella neglecta*
western porcupine grass, *Hesperostipa spartea*
western prairie fringed orchid, *Platanthera praeclara*
western sandpiper, *Calidris mauri*
western screech-owl, *Megascops kennicottii*
western silvery minnow, *Hybognathus argyritis*
western wheatgrass, *Pascopyrum smithii*
western wood lily, *Lilium philadelphicum*
western wood-pewee, *Contopus sordidulus*
wheat-stem sawfly, *Cephus cinctus*
white-footed mouse, *Peromyscus leucopus*
white spruce, *Picea glauca*
white-tailed deer, *Odocoileus virginianus*
white-tailed jackrabbit, *Lepus townsendii*
whooping crane, *Grus americana*
wild bergamot, *Monarda fistulosa*
wild turkey, *Meleagris gallopavo*
willet, *Catoptrophorus semipalmatus*
Wilson's phalarope, *Phalaropus tricolor*
wolf, *Canis lupus*
wolfberry, *Symphoricarpos occidentalis*
wolf willow, *Elaeagnus commutata*
wolverine, *Gulo gulo*
Wyoming ground squirrel, *Spermophilus elegans*

yarrow, *Achillea millefolium*
yellow-headed blackbird, *Xanthocephalus xanthocephalus*
yellow mud turtle, *Kinosternon flavescens*
yellow-shafted flicker, *Colaptes auratus*
yellow warbler, *Dendroica petechia*

zebra mussel, *Dreissena polymorpha*

Endemic Vertebrates of the Great Plains

The species listed below are entirely, or almost entirely, restricted to the Great Plains. Starred entries identify species that are considered to be vulnerable, threatened or endangered in part or all of their range. The information is derived from three sources: *Ecology and Conservation of Great Plains Vertebrates*, a book edited by Fritz L. Knopf and Fred B. Samson and published in 1996; *NatureServe Explorer* (www.natureserve.org/explorer/), October 2009 update, and the catalogue of species at risk issued by the Committee on the Status of Endangered Wildlife in Canada in 2010.

FISH

* Pallid sturgeon
* Western silvery minnow
* Sturgeon chub
* Sicklefin chub
  Redspot chub
* Red River shiner
  Wedgespot shiner
  Blacknose shiner
* Sharpnose shiner
* Kiamichi shiner
  Ozark shiner
* Topeka shiner
* Peppered shiner
* Smalleye shiner
* Arkansas River shiner
* Plateau shiner
  Duskystripe shiner
  Bleeding shiner
  Slim minnow
* Headwater catfish
* Widemouth blindcat
* Toothless blindcat
  Ozark madtom
  Checkered madtom
* Ouachita madtom
* Neosho madtom
* Caddo madtom
  Ozark cavefish

  Plains topminnow
  Ozark bass
* Arkansas darter
  Arkansas saddled darter
  Yoke darter
* Yellowcheek darter
* Niangua darter
* Paleback darter
  Stippled darter
  Orangebelly darter
  Missouri saddled darter
* Bluestripe darter
  Longnose darter
* Leopard darter
* Fountain darter
* Nueces roundnose minnow
* Clear creek gambusia
* Guadalupe bass

REPTILES

  Texas map turtle
  Ornate box turtle
* Dunes sagebrush lizard
* Prairie skink
* Harter's watersnake
  Plains garter snake
  Lined snake

AMPHIBIANS

* Cascade caverns salamander
* San Marcos salamander
* Texas salamander
* Comal blind salamander
* Barton springs salamander
* Jollyville plateau salamander
* Salado springs salamander
* Georgetown salamander
* Pedernales River springs salamander
* Edwards Plateau spring salamanders
* Blanco River spring salamander
* Texas blind salamander
* Comal springs salamander
* Blanco blind salamander
* Plains spadefoot toad
  Plains leopard frog

BIRDS

* Lesser prairie chicken
* Piping plover
* Mountain plover
* Long-billed curlew
  Marbled godwit
  Wilson's phalarope
  Franklin's gull
* Sprague's pipit
  Cassin's sparrow
  Baird's sparrow
  Lark bunting
* McCown's longspur
* Chestnut-collared longspur
* Black-capped vireo
* Golden-cheeked warbler
* Ferruginous hawk

MAMMALS

* Swift fox
* Black-footed ferret
  Spotted skunk
  Pronghorn
* Bison
* Elliot's short-tailed shrew
  Fringed myotis
* Texas kangaroo rat
* Black-tailed prairie dog
  Franklin's ground squirrel
  Richardson's ground squirrel
  Thirteen-lined ground squirrel
* Knox Jones's pocket gopher
* Llano pocket gopher
  Plains pocket gopher
  Olive-backed pocket mouse
  Plains pocket mouse
  Hispid pocket mouse
  Northern grasshopper mouse
  Plains harvest mouse
* Palo Duro mouse
  Prairie vole
  White-tailed jackrabbit

Map Credits

Unless otherwise noted below, the maps in this book are based on information in the public domain. All cartography is by the Canadian Plains Research Center at the University of Regina, except for the illustration on page 45, which is by Joan A. Williams, and the map on page 261, which is adapted from the Commission for Environmental Cooperation and the Nature Conservancy.

PAGE 5, Temperate Grasslands of the World, adapted from R.T. Coupland, ed., *Grassland Ecosystems of the World: Analysis of Grasslands and their Uses* (Cambridge: Cambridge University Press, 1979).

PAGE 6, Temperate Grasslands and Savannas of Canada and the United States, adapted from Taylor H. Ricketts, et al., *Terrestrial Ecoregions of North America: A Conservation Assessment* (Washington, D.C.: Island Press, 1999).

PAGE 9, Geography of the Great Plains, adapted from Antony R. Orme, ed., *The Physical Geography of North America* (New York: Oxford University Press, 2002).

PAGE 22, Ecoregions of the Great Plains, adapted from Taylor H. Ricketts, et al., *Terrestrial Ecoregions of North America: A Conservation Assessment* (Washington, D.C.: Island Press, 1999).

PAGE 45, Prairie Mountains, adapted from A.N. Strahler and J.H. Strahler, *Modern Physical Geography* (New York: John Wiley, 1978).

PAGE 73, Major Air Masses Affecting the Great Plains, adapted from Reid A. Bryson and F. Kenneth Hare, eds., *Climates of North America* (New York: Elsevier, 1974).

PAGE 80, Average Annual Precipitation on the Great Plains, data from ESRI ArcData Online, 2003.

PAGE 95, Soils of the Great Plains, adapted from Anthony Joern and Kathleen H. Keller, eds., *The Changing Prairie: North American Grasslands* (New York: Oxford University Press: 1995).

PAGE 108, Grassland Cover on the Great Plains, data from GeoInsight Corp.

PAGE 170 River Systems of the Great Plains, data from GeoInsight Corp.

PAGE 193, Woodlands and Savannas of the Great Plains Region, adapted from Taylor H. Ricketts, et al., *Terrestrial Ecoregions of North America: A Conservation Assessment* (Washington, D.C.: Island Press, 1999).

PAGE 226, Major Crops of the Great Plains, adapted from Frederick W. Boal and Stephen A. Royle, *North America: A Geographical Mosaic* (London: Arnold, 1999).

PAGE 261, Potential Areas for Large-Landscape Conservation on the Northern Great Plains, adapted from CEC and TNC, *North American Grassland Priority Conservation Areas: Technical Report and Documentation*, J.W. Karl and J. Hoth, eds.(Montreal: Commission for Environmental Cooperation and the Nature Conservancy, 2005).

PAGE 262, National and State/Provincial Parks on the Great Plains, data from Alberta Community Development, Saskatchewan Environment, and Manitoba Conservation, 2003, and University of California Santa Barbara Remote Sensing Research Unit, 1997.

PAGE 276, The Great Plains, data from ESRI Data and Maps Media Kit and GeoInsight Corp.

{ INDEX }

shorebirds; songbirds; woodpeckers; *individual species*

bison, 6, 7, 13–14, 15–16, 17, 24, 55, 58, 99, 106–7, 109, 123, 133, 149, 211; adaptations, 111; commercial production, 257; compared to cattle, 109–12, 114, 200; conservation, 256–58, 265; ecology, 117–19, 130–31, 135, 215, 226; evolution, 55, 57, 58–59, 109; hunters and hunting, 6, 7, 14–15, 17, 55, 106–7, 256; Kiowa myth, 106; migrations, 117–18, 119; original population, 17; wallows, 112

Bison antiquus, 55

Bison latifrons, 55

Bison priscus, 55

bison slaughter, 15, 17, 256

Bitterroot Range, Montana, 171

black bear, American, 54, 136, 196

black-billed magpie, 15, 200, 211

blackbirds, 112, 120, 147, 151, 215, 253, 255

black bullhead, 178

black crappie, 179

Black Elk, Nicholas, 185, 252

Blackfeet Reservation, Montana, 139, 259

black-footed ferret, 131, 264

Black Hills, 13, 32, 43, 195, 196, 236; geology and map, 45

Black Hills Coniferous Forest, 22, 23, 193

Black Hills redbelly snake, 264

Black Kettle National Grassland, Oklahoma, 145

Blackland Prairies, Texas, 22

blacknose shiner, 178

black-tailed jackrabbit, 141

black-tailed prairie dog, 99, 127, 129–30, 264. *See also* prairie dogs

black tern, 255

blister beetles, 238, 240, 243

blowout grass, 94

blowout penstemon, 260, 264

bluebunch wheatgrass, 118

blue grama, 20, 61, 63, 65, 68, 71, 83, 109, 113, 117, 131

bluegrasses, 235

blue-green algae, 30

blue jay, 194, 212

blue norther, 72

blue vervain, 124

blue-winged teal, 162, 255

bobcat, 54, 131, 141, 263

bobolink, 120, 122, 213, 254

Bombus griseocollis, 124

Bombus impatiens, 124

Bombus nevadensis, 124

boreal forest, 7, 8, 20, 164, 194, 195, 196, 217

Bovidae, or bovids, 107, 109, 133. *See also* bison; cattle/cattle industry

Bow River, 171

box elder, 202

Brachiosaurus, 41

braided rivers, 205–7

brassy minnow, 178

Brazeau Reservoir, Alberta, 178

Brazos River, 117, 171

Breeding Bird Survey, 122, 214, 254

broad-leafed plants. *See* forbs

brome grasses, 235; smooth brome, 116, 263

brown-headed cowbird, 215–16

brown-spotted range grasshopper, 236

buffalo. *See* bison

buffalo grass, 65, 69, 113, 117, 123, 131

Buffalo National Park, Alberta, 258

bufflehead, 164, 205

Bullock's oriole, 212, 213

bullsnakes, 168

bulrushes, 156

bumblebees, 124, 245

bunchgrass, 8

Burgess Shale, B.C., 37

bur oak, 18, 188, 191, 197

burrowing owl, 131, 132, 133, 264

butterflies and moths, 25–26, 43, 123, 124, 187, 229, 231, 267, 268. *See also individual species*

butterfly milkweed, 113, 114

butterfly weed, scarlet, 126

buttes, 54

C$_3$ and C$_4$ pathways, 71

cactus, 3, 67, 75, 89, 105, 121, 131

California Grasslands, 8

California gull, 237

Cambrian Period, 32

camels and camelids, 32, 37, 47, 49, 54, 55, 57, 58

Canada goose, 143

Canada thistle, 231

Canada wild rye, 233

Canadian prairies, 34, 73, 76, 77, 92, 128, 133, 153, 169, 171, 195, 209, 212. *See also* Alberta; Manitoba; Saskatchewan

Canadian River, 171

Canadian Shield. *See* Precambrian Shield

Canadian soil taxonomy, 94

canids, 138, 139. *See also* coyote; dire wolf; foxes; red fox; swift fox; wolf

canine distemper, 132

canvasback, 164, 255

Canyon Ferry Dam, Montana, 178

Caprock Escarpment, Texas, 9, 11, 195

Capulin Mountain, New Mexico, 11

Carabidae, 241. *See also* beetles; ground beetles

carbon dioxide, 35, 63, 65, 71, 198, 273

Carboniferous Period, 32, 36

cardinal, 211

Carlinian snapper, 222, 234, 235

carnivores. *See* canids; cats; grizzly bear; mesopredators; predator-prey relationships; skunks; weasels; *individual species*

carp, 178

Carpenter, David, 181

cats, 49, 54, 58. *See also* bobcat; cougar; lions

cat's claw mimosa, 191

cattails, 147, 153, 156, 158

cattle/cattle industry, 106, 109–11, 114, 119, 121, 130, 131, 136, 141, 265; and conservation, 259–66; and woodlands, 198, 200

cavitation, 186

cavity-nesting birds, 205

CEC (Commission on Environmental Cooperation), 253–54

cedar. *See* Ashe juniper; eastern red cedar; salt cedar

Central Kansas Wetlands, 151

Central Lowlands, 4, 9, 50

Central Rocky Mountains, 45

Central Tall Grasslands, 22, 23

Chahiksichahiks song, 14

Charles M. Russell National Wildlife Refuge, Montana, 259

chemical communication in plants, 240–41

chernozems, 94

chestnut-collared longspur, 121, 255

Cheyenne Bottoms, Kansas, 151

Cheyenne myth, 53

chinook salmon, 179

chinook winds, 25

chipping sparrow, 211

and trees, 189, 197; and wetlands, 156, 164, 166. *See also* Dirty Thirties
drought adaptations, 63–72, 78, 89; Rocky Mountain locust, 123; rodents, 127; soil organisms, 97
drought resistance in plants, 67
dry spells, 75, 77, 78. *See also* drought
duck factory, 153, 160, 168
ducks, 141, 151, 159, 162–69, 205, 246; coyote "control," 141
Ducks Unlimited, 167, 265
dung, 41, 119, 226
dung beetles, 99, 111
dwarf rhinoceros, 49

eared grebe, 165
early successional plants, 114. *See also* annual plants; community succession
earth-lodge people, 225
Earth's core, 30
Earth's crust, 30
earthworms, 97, 98, 101, 166
eastern bluebird, 212, 213
eastern hardwood/deciduous forest, 192, 212
eastern meadowlark, 213, 214
eastern red cedar, 197, 198
eastern spotted skunk, 216
eastern woodrat, 196
Echinacea, 124
ecological land classification and mapping, 19–20
ecological reserves. *See* conservation reserves
ecological traps, 248
ecology, 17–18, 19, 136
ecoregions, 17–20, 23–25, 80, 107; map, 22
ecosystem, 2, 4, 16, 17; defined, 14
ecotone, 194, 212, 216. *See also* Aspen Parklands; Prairie-and-Oak Transition; Southern Prairie-and-Oak Transition
ectothermic, 101, 237. *See also* cold-blooded
Edwards' hairstreak butterfly, 187
Edwards Plateau, 11, 264
Edwards Plateau Savannas, 22
effective precipitation, 79
elephants, Ice Age, 54, 55
elevation, 8
elk, 14, 23, 55, 192, 217, 218
Elk Island National Park, Alberta, 258
El Niño, 76–77

emergent vegetation, 156
Enchanted Rock, Texas, 32
endangered ecoregion, 107
endangered species. *See* species at risk
Endangered Species Act, U.S., 129, 194
endemic species, 264
Engelmann spruce, 196
Entomophaga grylli, 237
ephemeral wetlands, 154, 156, 163–64, 169. *See also* prairie potholes
erosion, 11, 13, 31, 32, 34, 39, 43, 45, 54, 106
eskers, 53
European corn borer, 231, 240
European starling, 247
evaporation and evaporative demand, 63, 77, 78, 79, 81, 149; trees and, 191, 194, 195
evergreen forest. *See* coniferous forest
evolution, 31, 35, 38, 136, 186, 234; evolutionary history, Great Plains, 29–59. *See also* adaptations to prairie environments; drought adaptations; extinctions and extirpations; subspecies
exotics, 116, 174, 179, 231
extinctions and extirpations, 23, 251, 256, 269; dinosaurs, 41, 42; Ice Age, 58, 133; Permian, 38–39; Rocky mountain locust, 123. *See also* species at risk
extrafloral nectaries, 102

Farm Bill, 2009, U.S., 271
farming. *See* agroecosystems; cropland; farming, history of; herbicides; insecticides; organic, or holistic, farming; settlers and settlement era
farming, history of, 2, 5, 15, 176, 225–31, 244–46; Green Revolution, 269–72, 273–75; income-support programs, 271; no-till and minimum-till methods, 247–48, 272; organic methods, 248, 272
farming-and-hunting cultures, 5, 14, 55, 199, 225
fences and fencerows, 16, 114, 117, 121, 229, 240, 247, 271
Ferril, Thomas Hornsby, 143
ferruginous hawk, 143, 255, 264, 270
fescue grassland, 20
fescues, 235
field margins, 243, 246. *See also* roadsides and ditches

fire, 14, 76, 81, 114, 117, 120, 196; suppression, 120, 194, 198, 200, 207, 226; and trees, 189, 197
First Peoples. *See* Native people
fish, 35, 172, 173, 174, 175–76, 178, 179, 180, 264; adaptations to prairie rivers, 180
Fishing in the West, 181
flathead chub, 181
Flathead Reservation, Montana, 256
flat-topped aster, 233
flies, 122, 136, 151, 238, 240
Flint Hills, Kansas, 45, 98, 107, 114, 197, 267
Flint Hills Tall Grasslands, 22
floodplain, 174–75, 200
flood-pulse cycle, 175–76, 178, 180, 202, 205
floods and flood control, 76, 179, 202, 207
floodwater mosquitoes, 158
Foothills Grasslands, 20, 22
forbs, 63, 186, 195, 234. *See also* individual species
forest, 4, 46, 54, 68, 89. *See also* boreal forest; coniferous forest; eastern hardwood/deciduous forest; savannas; scarp woodlands; woodlands
Formica integra, 187
Fort Peck Dam, Montana, 178
Fort Randall Dam, South Dakota, 178
fossil trackway, 37, 47
foxes, 16, 131, 139–41, 168, 216, 229, 247
fox squirrel, 216
fragmentation, 246, 253
Franklin's ground squirrel, 128, 216
Frenchman River, 171
Frenchman Valley, 41, 42, 105, 106, 107
freshwater clams, 182
freshwater shrimp, 151
fringed sagewort, 88
frogs and toads, 39, 43, 151, 161, 166, 173, 174, 176, 216, 229; status, 175
Front Range, Colorado, 54
frost wedging, 96
fungi, 86, 87, 91, 99, 238

gadwall, 163, 255
galleta, 65
gardening cultures, 6, 14, 199, 225
Gardiner Dam, Saskatchewan, 178
Garrison Dam, North Dakota, 178, 201
garter snakes, 166, 174
Gavins Point Dam, 178

THE DAVID SUZUKI FOUNDATION

The David Suzuki Foundation works through science and education to protect the diversity of nature and our quality of life, now and for the future.

With a goal of achieving sustainability within a generation, the Foundation collaborates with scientists, business and industry, academia, government and non-governmental organizations. We seek the best research to provide innovative solutions that will help build a clean, competitive economy that does not threaten the natural services that support all life.

The Foundation is a federally registered independent charity that is supported with the help of over 50,000 individual donors across Canada and around the world.

We invite you to become a member. For more information on how you can support our work, please contact us:

The David Suzuki Foundation
219–2211 West 4th Avenue
Vancouver, BC
Canada V6K 4S2
www.davidsuzuki.org
contact@davidsuzuki.org
Tel: 604-732-4228
Fax: 604-732-0752

Checks can be made payable to The David Suzuki Foundation.
All donations are tax-deductible.

Canadian charitable registration: (BN) 12775 6716 RR0001

U.S. charitable registration: #94-3204049